FUNDAMENTALS OF PLANT PATHOLOGY

Edited by
Arit Efretuei

ArclerPress

Fundamentals of Plant Pathology

Edited by Arit Efretuei

Edition 2018

ArclerPress

Published by
Arcler Press LLC
708 3rd Avenue, 6th Floor
New York
NY 10017
United States of America
www.arclerpress.com

ISBN: 978-1-68094-557-7 (Hardback)

Printed in United States of America

Library of Congress Control Number: 2017933128

About the Editor

Arit Efretuei

Arit Efretuei obtained her PhD in Agriculture from the University of Reading, UK in 2015. Her interests are in crop nutrient uptake, soil nutrient management and mitigation of soil nutrient losses to the environment. She is presently working as a freelance agronomic writer.

Table of Contents

2 PLANT DISEASE EPIDEMIOLOGY

3 PLANT VIRUSES

4 PHYSIOLOGICAL PLANT DISORDER

6 GENERAL CHARACTERISTICS OF VIROIDS

7 PHYTOPLASMA

8 FUNGUS

List of Abbreviations

ATP	Adenosine Triphosphate
ASBV	Avocado Sunblotch Viroid
CNBM	Carbon Nutrient Balance Model
CaMV	Cauliflower Mosaic Virus
CCCVd	Coconut cadang-cadang viroid
CTiVd	Coconut Tinangaja Viroid
DAMP	Damage-Associated Compounds
dsRNA	Double-Stranded RNA
ETI	Effector Triggered Immunity
EPIPRE	EPIdemiology, PREdiction, and PREvention
EPPO	European and Mediterranean Plant Protection Organization
HIPVs	Herbivore-Induced Plant Volatiles
HPR	Host-Plant Resistance
INSV	Impatiens Necrotic Spot Virus
ICP	Inductively Coupled Plasma Spectrometry
LOX	Lipoxygenase
mRNA	Messenger RNA
MFOs	Mixed Function Oxidases
MLOs	Mycoplasma-Like Organisms
PAMPs or MAMPs	Pathogen Or Microbial–Associated Molecular Patterns
PTI	Pattern-Triggered Immunity
PMMoV	Pepper Mild Mottle Virus

PEPC	Phosphoenolpyruvate Carboxylase
PPLO	Pleuropneumonia-Like Organisms
PEG	polyethylene glycol
PSTVd	Potato Spindle Tuber Viroid
RCC	Renal Cell Carcinoma
ssRNA	Single Stranded RNA
snRNA	Small Nuclear RNA
TMV	Tobacco Mosaic Virus
ToMV	Tomato Mosaic Virus
TSWV	Tomato Spotted Wilt Virus
TAL	Transcription Activator–Like

Preface

Plant Pathology is defined as the study of the organisms and environmental conditions that cause diseases in plants, the mechanisms by which this occurs, the interactions between these causal agents and the plant, and the methods of managing or controlling plant disease. It also draws on knowledge from other scientific fields such as mycology, microbiology, virology, biochemistry, bio-informatics, etc. Plant diseases are caused by biotic agents like fungi, bacteria, actinomycets, Mycoplasma, viruses, nematodes, and flowering parasites or by abiotic like unfavorable environmental conditions or nutritional deficiencies. Plant diseases are known from times preceding the earliest writings. Fossil evidence indicates that plants were affected by disease 250 million years ago. Loss of crops from plant diseases may result in hunger and starvation, especially in less developed countries where access to disease-control methods is limited and annual losses of 30 to 50 percent are common for major crops. In some years, losses are much greater, producing catastrophic results for those who depend on the crop for food. Major disease outbreaks among food crops have led to famines and mass migrations throughout history. Losses from plant diseases also can have a significant economic impact, causing a reduction in income for crop producers and distributors and higher prices for consumers. Plant diseases are a normal part of nature and one of many ecological factors that help keep the hundreds of thousands of living plants and animals in balance with one another. Plant cells contain special signaling pathways that enhance their defenses against insects, animals, and pathogens. Plant diseases can be broadly classified according to the nature of their primary causal agent, either infectious or noninfectious. Infectious plant diseases are caused by a pathogenic organism such as a fungus, bacterium, mycoplasma, virus, viroid, nematode, or parasitic flowering plant. An infectious agent is capable of reproducing within or on its host and spreading from one susceptible host to another. Noninfectious plant diseases are caused by unfavorable growing conditions, including extremes of temperature, disadvantageous relationships between moisture and oxygen, toxic substances in the soil or atmosphere, and an excess or deficiency of an essential mineral. Because noninfectious causal agents are not organisms capable of reproducing within a host, they are not transmissible.

Content Coverage

Chapter One presents the scientific study of diseases in plants caused by pathogens (infectious organisms) and environmental conditions (physiological factors).

Chapter Two focuses on plant disease epidemiology, which is the study of disease in plant populations. Much like diseases of humans and other animals, plant diseases occur due to pathogens such as bacteria, viruses, fungi, oomycetes, nematodes, phytoplasmas, protozoa, and parasitic plants.

Chapter Three focuses on plant viruses that affect plants. Like all other viruses, plant viruses are obligate intracellular parasites that do not have the molecular machinery to replicate without a host. Plant viruses are pathogenic to higher plants.

Chapter Four focuses on physiological plant disorders caused by non-pathological conditions such as poor light, adverse weather, water-logging, phytotoxic compounds or a lack of nutrients, and affect the functioning of the plant system.

Chapter Five is concerned plant defense against herbivory or host-plant resistance (HPR) that describes a range of adaptations evolved by plants which improve their survival and reproduction by reducing the impact of herbivores.

Chapter Six presents an emphasis on viroids that are the smallest infectious pathogens known, larger only than prions, which, as nucleic acid-free, misfolded protein molecules, only whose abnormal conformation is transmitted, are not strictly infectious pathogens.

Chapter Seven focuses on the role of phytoplasmas that are obligate parasites of plant phloem tissue and transmitting insects. Phytoplasmas are pathogens of agriculturally important plants, including coconut, sugarcane, and sandalwood, causing a wide variety of symptoms that range from mild yellowing to death of infected plants

Chapter Eight focuses on the fungus that includes unicellular microorganisms such as yeasts and molds, as well as multicellular fungi that produce familiar fruiting forms known as mushrooms. Many fungi are parasites on plants, animals (including humans), and other fungi. Serious pathogens of many cultivated plants causing extensive damage and losses to agriculture and forestry include the rice blast fungus Magnaporthe oryzae, tree pathogens such as Ophiostoma ulmi and Ophiostoma novo-ulmi causing Dutch elm disease, and Cryphonectria parasitica responsible for chestnut blight, and plant pathogens in the genera Fusarium, Ustilago, Alternaria, and Cochliobolus.

CHAPTER 1

Introduction to Plant Pathology

Chapter Contents

The science of plant diseases is called plant pathology, or phytopathology. There are a wide variety of microorganisms such as fungi, bacteria, viruses and nematodes that cause these diseases. Diseases caused by these pathogens are often called biotic diseases. Also, environmental conditions, such as winter damage or drought stress, can cause disease in plants. Diseases caused by these factors are often called abiotic diseases.

INTRODUCTION

Plant pathology (also phytopathology) is the scientific study of diseases in plants caused by pathogens (infectious organisms) and environmental conditions (physiological factors). Organisms that cause infectious disease include fungi, oomycetes, bacteria, viruses, viroids, virus-like organisms, phytoplasmas, protozoa, nematodes and parasitic plants. Not included are ectoparasites like insects, mites, vertebrate, or other pests that affect plant health by consumption of plant tissues. Plant pathology also involves the study of pathogen identification, disease etiology, disease cycles, economic impact, plant disease epidemiology, plant disease resistance, how plant diseases affect humans and animals, pathosystem genetics, and management of plant diseases.

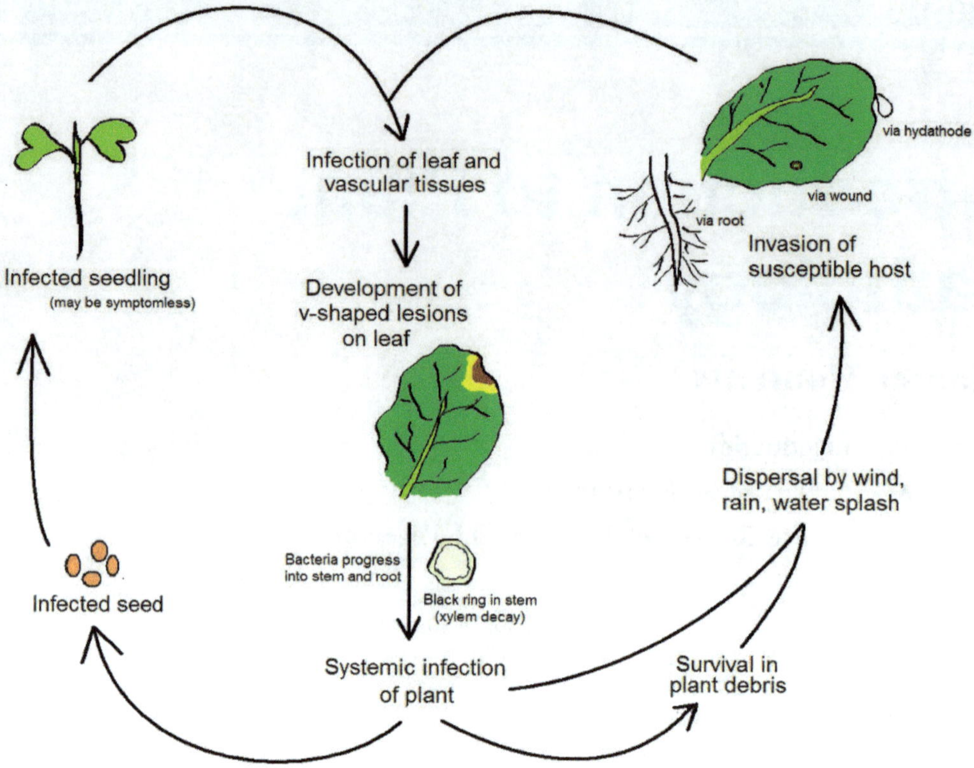

Life cycle of the black rot pathogen, Xanthomonas campestris pathovar campes

Overview

Control of plant diseases is crucial to the reliable production of food, and it provides significant reductions in agricultural use of land, water, fuel and other inputs. Plants in both natural and cultivated populations carry inherent disease resistance, but there are numerous examples of devastating plant disease impacts, as well as recurrent severe plant diseases (see rice blast, soybean cyst nematode, citrus canker). However, disease control is reasonably successful for most crops. Disease control is achieved by use of plants that have been bred for good resistance to many diseases, and by plant cultivation approaches such as crop rotation, use of pathogen-free seed, appropriate planting date and plant density, control of field moisture, and pesticide use. Across large regions and many crop species. Continuing advances in the science of plant pathology are needed to improve disease control, and to keep up with changes in disease pressure caused by the ongoing evolution and movement of plant pathogens and by changes in agricultural practices. Plant diseases cause major economic losses for farmers worldwide. The Food and Agriculture Organization estimates indeed that pests and diseases are responsible for about 25% of crop loss. To solve this issue, new methods are needed to detect diseases and pests early, such as novel sensors that detect plant odours and spectroscopy and biophotonics that are able to diagnostic plant health and metabolism.

Plant Pathogens

Powdery mildew, a biotrophic fungus

Fungi

Most phytopathogenic fungi belong to the Ascomycetes and the Basidiomycetes.

The fungi reproduce both sexually and asexually via the production of spores and other structures. Spores may be spread long distances by air or water, or they may be soilborne. Many soil inhabiting fungi are capable of living saprotrophically, carrying out the part of their life cycle in the soil. These are known as facultative saprotrophs.

Fungal diseases may be controlled through the use of fungicides and other agriculture practices. However, new races of fungi often evolve that are resistant to various fungicides.

Biotrophic fungal pathogens colonize living plant tissue and obtain nutrients from living host cells. Necrotrophic fungal pathogens infect and kill host tissue and extract nutrients from the dead host cells. See the powdery mildew and rice blast images, below.

Rice blast, caused by a necrotrophic fungus

Significant fungal plant pathogens include:

Ascomycetes

- *Fusarium* spp. (causal agents of Fusarium wilt disease)
- *Thielaviopsis* spp. (causal agents of: canker rot, black root rot, *Thielaviopsis* root rot)
- *Verticillium* spp.
- *Magnaporthe grisea* (causal agent of rice blast)
- *Sclerotinia sclerotiorum* (causal agent of cottony rot)

Basidiomycetes

- *Ustilago* spp. (causal agents of smut)
- *Rhizoctonia* spp.
- *Phakospora pachyrhizi* (causal agent of soybean rust)
- *Puccinia* spp. (causal agents of severe rusts of virtually all cereal grains and cultivated grasses)
- *Armillaria* spp. (the so-called honey fungus species, which are virulent pathogens of trees and produce edible mushrooms)

Fungus-Like Organisms

Oomycetes

The oomycetes are not true fungi but are fungus-like organisms. They include some of the most destructive plant pathogens including the genus*Phytophthora*, which includes the causal agents of potato late blight and sudden oak death. Particular species of oomycetes are responsible for root rot. Despite not being closely related to the fungi, the oomycetes have developed very similar infection strategies. Oomycetes are capable of using effector proteins to turn off a plant's defenses in its infection process. Plant pathologists commonly group them with fungal pathogens.

Significant oomycete plant pathogens

- *Pythium* spp.
- *Phytophthora* spp., including the causal agent of the Great Irish Famine (1845–1849)

Phytomyxea

Some slime molds in Phytomyxea cause important diseases, including club root in cabbage and its relatives and powdery scab in potatoes. These are caused by species of *Plasmodiophora* and *Spongospora*, respectively.

Bacteria

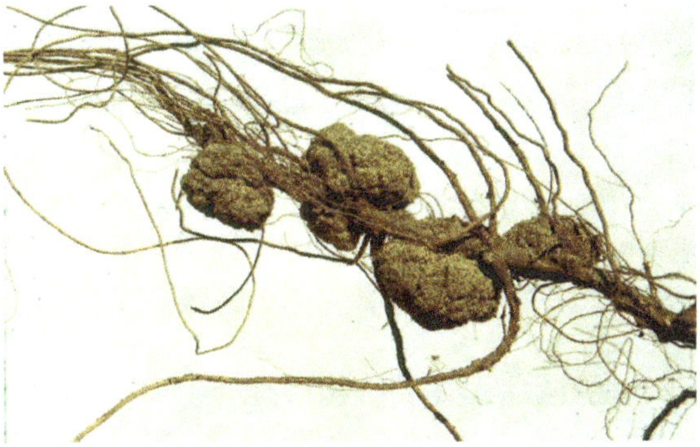

Crown gall disease caused by Agrobacterium

Most bacteria that are associated with plants are actually saprotrophic and do no harm to the plant itself. However, a small number, around 100 known species, are able to cause disease. Bacterial diseases are much more prevalent in subtropical and tropical regions of the world.

Most plant pathogenic bacteria are rod-shaped (bacilli). In order to be able to colonize the plant they have specific pathogenicity factors. Five main types of bacterial pathogenicity factors are known: uses of cell wall–degrading enzymes, toxins, effector proteins, phytohormones and exopolysaccharides.

Pathogens such as *Erwinia* species use cell wall–degrading enzymes to cause soft rot. *Agrobacterium* species change the level of auxins to cause tumours with phytohormones. Exopolysaccharides are produced by bacteria and block xylem vessels, often leading to the death of the plant.

Bacteria control the production of pathogenicity factors via quorum sensing.

Vitis vinifera with "Ca. Phytoplasma vitis" infection

Significant bacterial plant pathogens:

- Burkholderia
- Proteobacteria
  - *Xanthomonas* spp.
  - *Pseudomonas* spp.
- Pseudomonas syringae pv. tomato causes tomato plants to produce less fruit, and it "continues to adapt to the tomato by minimizing its recognition by the tomato immune system."

Phytoplasmas ('Mycoplasma-Like Organisms') and Spiroplasmas

Phytoplasma and *Spiroplasma* are a genre of bacteria that lack cell walls and are related to the mycoplasmas, which are human pathogens. Together they are referred to as the mollicutes. They also tend to have smaller genomes than most other bacteria. They are normally transmitted by sap-sucking insects, being transferred into the plants phloem where it reproduces.

Tobacco mosaic virus

Viruses, Viroids and Virus-Like Organisms

There are many types of plant virus, and some are even asymptomatic. Under normal circumstances, plant viruses cause only a loss of crop yield. Therefore, it is not economically viable to try to control them, the exception being when they infect perennial species, such as fruit trees.

Most plant viruses have small, single-stranded RNA genomes. However some plant viruses also have double stranded RNA or single or double stranded DNA genomes. These genomes may encode only three or four proteins: a replicase, a coat protein, a movement protein, in order to allow cell to cell movement through plasmodesmata, and sometimes a protein that allows transmission by a vector. Plant viruses can have several more proteins and employ many different molecular translation methods.

Plant viruses are generally transmitted from plant to plant by a vector, but mechanical and seed transmission also occur. Vector transmission is often by an insect (for example, aphids), but some fungi, nematodes, and protozoa have been shown to be viral vectors. In many cases, the insect and virus are specific for virus transmission such as the beet leafhopper that transmits the curly top virus causing disease in several crop plants.

Nematodes

Root-knot nematode galls

Nematodes are small, multicellular wormlike animals. Many live freely in the soil, but there are some species that parasitize plant roots. They are a problem in tropical and subtropical regions of the world, where they may infect crops. Potato cyst nematodes (*Globodera pallida* and *G. rostochiensis*) are widely distributed in Europe and North and South America and cause $300 million worth of damage in Europe every year. Root knot nematodes have quite a large host range, they parasitize plant root systems and thus directly affect the uptake of water and nutrients needed for normal plant growth and reproduction, whereas cyst nematodes tend to be able to infect only a few species. Nematodes are able to cause radical changes in root cells in order to facilitate their lifestyle.

Protozoa and Algae

There are a few examples of plant diseases caused by protozoa (e.g., *Phytomonas*, a kinetoplastid). They are transmitted as zoospores that are very durable, and may be able to survive in a resting state in the soil for many years. They have also been shown to transmit plant viruses.

When the motile zoospores come into contact with a root hair they produce a plasmodium and invade the roots.

Some colourless parasitic algae (e.g., *Cephaleuros*) also cause plant diseases.

Parasitic Plants

Parasitic plants such as mistletoe and dodder are included in the study of phytopathology. Dodder, for example, is used as a conduit either for the transmission of viruses or virus-like agents from a host plant to a plant that is not typically a host or for an agent that is not graft-transmissible.

Common Pathogenic Infection Methods

- Cell wall-degrading enzymes: These are used to break down the plant cell wall in order to release the nutrients inside.
- Toxins: These can be non-host-specific, which damage all plants, or host-specific, which cause damage only on a host plant.
- Effector proteins: These can be secreted into the extracellular environment or directly into the host cell, often via the Type three secretion system. Some effectors are known to suppress host defense processes. This can include: reducing the plants internal signaling mechanisms or reduction of phytochemicals production. Bacteria, fungus and oomycetes are known for this function.

Physiological Plant Disorders

Significant abiotic disorders can be caused by:

Natural

Drought

- Frost damage and breakage by snow and hail
- Flooding and poor drainage
- Nutrient deficiency
- Salt deposition and other soluble mineral excesses (e.g., gypsum)
- Wind (windburn and breakage by hurricanes and tornadoes)
- Lightning and wildfire (also often man-made)
- ***Man-made*** (arguably not abiotic, but usually regarded as such)
- Soil compaction
- Pollution of air, soil, or both
- Salt from winter road salt application or irrigation
- Herbicide over-application
- Poor education and training of people working with plants (e.g. lawnmower damage to trees)
- Vandalism

Epidemiology

The study of phytopathology epidemics

Management

Quarantine

A diseased patch of vegetation or individual plants can be isolated from other, healthy growth. Specimens may be destroyed or relocated into a greenhouse for treatment or study. Another option is to avoid the introduction of harmful nonnative organisms by controlling all human traffic and activity (e.g., AQIS), although legislation and enforcement are crucial in order to ensure lasting effectiveness.

Cultural

Farming in some societies is kept on a small scale, tended by peoples whose culture includes farming traditions going back to ancient times. (An example of such traditions would be lifelong training in techniques of plot terracing, weather anticipation and response, fertilization, grafting, seed care, and dedicated gardening.) Plants that are intently monitored often benefit from not only active external protection but also a greater overall vigor. While primitive in the sense of being the most labor-intensive solution by far, where practical or necessary it is more than adequate.

Plant Resistance

Sophisticated agricultural developments now allow growers to choose from among systematically cross-bred species to ensure the greatest hardiness in their crops, as suited for a particular region's pathological profile. Breeding practices have been perfected over centuries, but with the advent of genetic manipulation even finer control of a crop's immunity traits is possible. The engineering of food plants may be less rewarding, however, as higher output is frequently offset by popular suspicion and negative opinion about this "tampering" with nature.

Chemical

Many natural and synthetic compounds can be employed to combat the above threats. This method works by directly eliminating disease-causing organisms or curbing their spread; however, it has been shown to have too broad an effect, typically, to be good for the local ecosystem. From an economic standpoint, all but the simplest natural additives may disqualify a product from "organic" status, potentially reducing the value of the yield.

Biological

Crop rotation may be an effective means to prevent a parasitic population from becoming well-established, as an organism affecting leaves would be starved when the leafy crop is replaced by a tuberous type, etc. Other means to undermine parasites without attacking them directly may exist.

Integrated

The use of two or more of these methods in combination offers a higher chance of effectiveness.

Timeline of Plant Pathology

- 300–286 BC Theophrastus, father of botany, wrote and studied diseases of trees, cereals and legumes
- 1665 Robert Hooke illustrates a plant-pathogenic fungal disease, rose rust
- 1675 Anton van Leeuwenhouek invents the compound microscope, in 1683 describes bacteria seen with the microscope
- 1729 Pier Antonio Micheli, father of mycology, observes spores for the first time, conducts germination experiments
- 1755 Tillet reports on treatment of seeds
- 1802 Lime sulfur first used to control plant disease
- 1845–1849 Potato late blight epidemic in Ireland
- 1853 Heinrich Anton de Bary father of modern mycology, establishes that fungi are the cause, not the result, of plant diseases, publishes "Untersuchungen uber die Brandpilze"
- 1858 Julius Kühn publishes "Die Krankheiten der Kultergewachse"
- 1865 M. Planchon discovers a new species of *Phylloxera*, which was named *Phylloxera vastatrix.*
- 1868–1882 Coffee rust epidemic in Sri Lanka
- 1875 Mikhail Woronin identified the cause of clubroot as a "plasmodiophorous organism" and gave it the name *Plasmodiophora brassicae*
- 1876 *Fusarium oxysporum* f.sp. *cubense*, responsible for Panama disease, discovered in bananas in Australia
- 1878–1885 Downy mildew of grape epidemic in France
- 1879 Robert Koch establishes germ theory: diseases are caused by microorganisms
- 1882 *Lehrbuch der Baumkrankheiten* (*Textbook of Diseases of Trees*), by Robert Hartig, is published in Berlin, the first textbook of forest pathology.
- 1885 Bordeaux mixture introduced by Pierre-Marie-Alexis Millardet to control downy mildew on grape
- 1885 Experimental proof that bacteria can cause plant diseases: "Erwinia amylovora" and fire blight of apple
- 1886–1898 Recognition of plant viral diseases: Tobacco mosaic virus
- 1889 Introduction of hot water treatment of seed for disease control by Jensen
- 1902 First chair of plant pathology established, in Copenhagen
- 1904 Mendelian inheritance of cereal rust resistance demonstrated

- 1907 First academic department of plant pathology established, at Cornell University
- 1908 American Phytopathological Society founded
- 1910 Panama disease reaches Western Hemisphere
- 1911 Scientific journal *Phytopathology* founded
- 1925 Panama disease reaches every banana-growing country in the Western Hemisphere
- 1951 European and Mediterranean Plant Protection Organization (EPPO) founded
- 1967 Recognition of plant pathogenic mycoplasma-like organisms
- 1971 T. O. Diener discovers viroids, organisms smaller than viruses

PLANT DISEASE RESISTANCE

Plant disease resistance protects plants from pathogens in two ways: by pre-formed structures and chemicals, and by infection-induced responses of the immune system. Relative to a susceptible plant, disease resistance is the reduction of pathogen growth on or in the plant, while the term disease tolerance describes plants that exhibit little disease damage despite substantial pathogen levels. Disease outcome is determined by the three-way interaction of the pathogen, the plant and the environmental conditions (an interaction known as the disease triangle).

Defense-activating compounds can move cell-to-cell and systemically through the plant vascular system. However, plants do not have circulating immune cells, so most cell types exhibit a broad suite of antimicrobial defenses. Although obvious qualitative differences in disease resistance can be observed when multiple specimens are compared (allowing classification as "resistant" or "susceptible" after infection by the same pathogen strain at similar inoculum levels in similar environments), a gradation of quantitative differences in disease resistance is more typically observed between plant strains or genotypes. Plants consistently resist certain pathogens but succumb to others; resistance is usually specific to certain pathogen species or pathogen strains.

Cankers caused by Chestnut blight, a disease that affects the chestnut tree

Background

Plant disease resistance is crucial to the reliable production of food, and it provides significant reductions in agricultural use of land, water, fuel and other inputs. Plants in both natural and cultivated populations carry inherent disease resistance, but this has not always protected them.

The late blight Irish potato famine of the 1840s was caused by the oomycete Phytophthora infestans. The world's first mass-cultivated banana cultivar Gros Michel was lost in the 1920s to Panama disease caused by the fungus Fusarium oxysporum. The current wheat stem, leaf, and yellow stripe rust epidemics spreading from East Africa into the Indian subcontinent are caused by rust fungi Puccinia graminis and P. striiformis. Other epidemics include Chestnut blight, as well as recurrent severe plant diseases such as Rice blast, Soybean cyst nematode, Citrus canker.

Plant pathogens can spread rapidly over great distances, vectored by water, wind, insects, and humans. Across large regions and many crop species, it is estimated that diseases typically reduce plant yields by 10% every year in more developed nations or agricultural systems, but yield loss to diseases often exceeds 20% in less developed settings, an estimated 15% of global crop production.

However, disease control is reasonably successful for most crops. Disease control is achieved by use of plants that have been bred for good resistance to many diseases, and by plant cultivation approaches such as crop rotation, pathogen-free seed, appropriate planting date and plant density, control of field moisture and pesticide use.

Common Disease Resistance Mechanisms

Pre-Formed Structures and Compounds

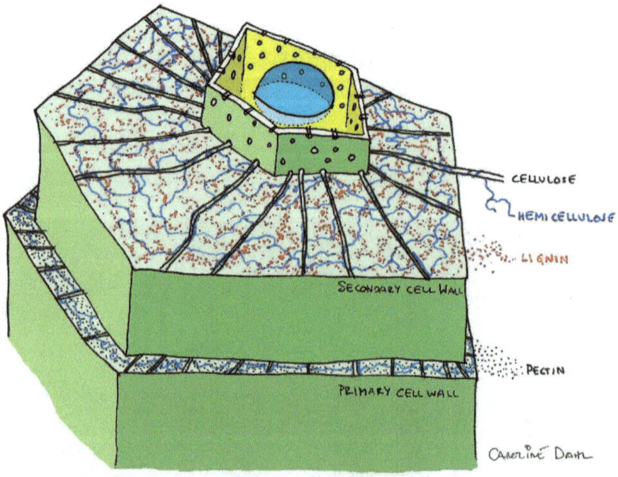

secondary plant wall

- Plant cuticle/surface
- Plant cell walls
- Antimicrobial chemicals (for example: glucosides, saponins)
- Antimicrobial proteins
- Enzyme inhibitors
- Detoxifying enzymes that break down pathogen-derived toxins
- Receptors that perceive pathogen presence and activate inducible plant defences

Inducible Post-Infection Plant Defenses

- Cell wall reinforcement (callose, lignin, suberin, cell wall proteins)
- Antimicrobial chemicals, including reactive oxygen species such as hydrogen peroxide or peroxynitrite, or more complex phytoalexins such as genistein or camalexin
- Antimicrobial proteins such as defensins, thionins, or PR-1
- Antimicrobial enzymes such as chitinases, beta-glucanases, or peroxidases
- Hypersensitive response - a rapid host cell death response associated with defence mediated by "Resistance genes."(Bryant, Tracy 2008).

Immune System

The plant immune system carries two interconnected tiers of receptors, one most frequently sensing molecules outside the cell and the other most frequently sensing molecules inside the cell. Both systems sense the intruder, respond to the intrusion and signal to the rest of the plant and sometimes to neighboring plants that the intruder is present. The two systems detect different types of pathogen molecules and classes of plant receptor proteins.

The first tier is primarily governed by pattern recognition receptors that are activated by recognition of evolutionarily conserved pathogen or microbial–associated molecular patterns (PAMPs or MAMPs). Activation of PRRs leads to intracellular signaling, transcriptional reprogramming, and biosynthesis of a complex output response that limits colonization. The system is known as PAMP-Triggered Immunity or as Pattern-Triggered Immunity (PTI).

The second tier, primarily governed by R gene products, is often termed effector-triggered immunity (ETI). ETI is typically activated by the presence of specific pathogen "effectors" and then triggers strong antimicrobial responses (see R gene section below).

In addition to PTI and ETI, plant defenses can be activated by the sensing of damage-associated compounds (DAMP), such as portions of the plant cell wall released during pathogenic infection.

Responses activated by PTI and ETI receptors include ion channel gating, oxidative burst, cellular redox changes, or protein kinase cascades that directly activate cellular changes (such as cell wall reinforcement or antimicrobial production), or activate changes in gene expression that then elevate other defensive responses

Plant immune systems show some mechanistic similarities with the immune systems of insects and mammals, but also exhibit many plant-specific characteristics. The two above-described tiers are central to plant immunity but do not fully describe plant immune systems. In addition, many specific examples of apparent PTI or ETI violate common PTI/ETI definitions, suggesting a need for broadened definitions and/or paradigms.

Pamp-Triggered Immunity

PAMPs, conserved molecules that inhabit multiple pathogen genera, are referred to as MAMPs by many researchers. The defenses induced by MAMP perception are sufficient to repel most pathogens. However, pathogen effector proteins (see below) are adapted to suppress basal defenses such as PTI. Many receptors for MAMPs (and DAMPs) have been discovered. MAMPs and DAMPs are often detected by transmembrane receptor-kinases that carry LRR or LysM extracellular domains.

Effector Triggered Immunity

Effector Triggered Immunity (ETI) is activated by the presence of pathogen effectors. The ETI response is reliant on R genes, and is activated by specific pathogen strains. Plant ETI often causes an apoptotic hypersensitive response.

R Genes and R Proteins

Plants have evolved R genes (resistance genes) whose products mediate resistance to specific virus, bacteria, oomycete, fungus, nematode or insect strains. R gene products are proteins that allow recognition of specific pathogen effectors, either through direct binding or by recognition of the effector's alteration of a host protein. Many R genes encode NB-LRR proteins (proteins with nucleotide-binding and leucine-rich repeat domains, also known as NLR proteins or STAND proteins, among other names). Most plant immune systems carry a repertoire of 100-600 different R gene homologs. Individual R genes have been demonstrated to mediate resistance to specific virus, bacteria, oomycete, fungus, nematode or insect strains. R gene products control a broad set of disease resistance responses whose induction is often sufficient to stop further pathogen growth/spread.

Studied R genes usually confer specificity for particular strains of a pathogen species (those that express the recognized effector). As first noted by Harold Flor in his mid-20th century formulation of the gene-for-gene relationship, a plant R gene has specificity for a pathogen avirulence gene (Avr gene). Avirulence genes are now known to encode effectors. The pathogen Avr gene must have matched specificity with the R gene for that R gene to confer resistance, suggesting a receptor/ligand interaction for Avr and R genes. Alternatively, an effector can modify its host cellular target (or a molecular decoy of that target), and the R gene product (NLR protein) activates defenses when it detects the modified form of the target or decoy.

Effector Biology

Effectors are central to the pathogenic or symbiotic potential of microbes and microscopic plant-colonizing animals such as nematodes.Effectors typically are proteins that are delivered mostly outside the microbe and into the host cell. Effectors manipulate cell physiology and development. As such, effectors offer examples of co-evolution (example: a fungal protein that functions outside of the fungus but inside of plant cells has evolved to take on plant-specific functions). Pathogen host range is determined, among other things, by the presence of appropriate effectors that allow colonization of a particular host. Pathogen-derived effectors are a powerful tool to identify host functions that are important in disease. Apparently most effectors function to manipulate host physiology to allow disease to occur. Well-studied bacterial plant pathogens typically express a few dozen effectors, often delivered into the host by a Type III secretion apparatus. Fungal, oomycete and nematode plant pathogens apparently express a few hundred effectors.

So-called "core" effectors are defined operationally by their wide distribution across the population of a particular pathogen and their substantial contribution to pathogen virulence. Genomics can be used to identify core effectors, which can then be used to discover new R gene alleles, which can be used in plant breeding for disease resistance.

RNA Silencing and Systemic Acquired Resistance Elicited by Prior Infections

Against viruses, plants often induce pathogen-specific gene silencing mechanisms mediated by RNA interference. This is a form of adaptive immunity.

Plant immune systems also can respond to an initial infection in one part of the plant by physiologically elevating the capacity for a successful defense response in other parts. Such responses include systemic acquired resistance, largely mediated by salicylic acid-dependent pathways, and induced systemic resistance, largely mediated by jasmonic acid-dependent pathways.

Species-Level Resistance

In a small number of cases, plant genes are effective against an entire pathogen species, even though that species that is pathogenic on other genotypes of that host species. Examples include barley MLO against powdery mildew, wheat Lr34 against leaf rust and wheat Yr36 against stripe rust. An array of mechanisms for this type of resistance may exist depending on the particular gene and plant-pathogen combination. Other reasons for effective plant immunity can include a lack of coadaptation (the pathogen and/or plant lack multiple mechanisms needed for colonization and growth within that host species), or a particularly effective suite of pre-formed defenses.

Signaling Mechanisms

Perception of Pathogen Presence

Plant defense signaling is activated by the pathogen-detecting receptors that are described in an above section. The activated receptors frequently elicit reactive oxygen and nitric oxide production, calcium, potassium and proton ion fluxes, altered levels of salicylic acid and other hormones and activation of MAP kinases and other specific protein kinases. These events in turn typically lead to the modification of proteins that control gene transcription, and the activation of defense-associated gene expression.

Transcription Factors and the Hormone Response

Numerous genes and/or proteins as well as other molecules have been identified that mediate plant defense signal transduction.Cytoskeleton and vesicle trafficking dynamics help to orient plant defense responses toward the point of pathogen attack.

Mechanisms of Transcription Factors And Hormones

Plant immune system activity is regulated in part by signaling hormones such as:

- Salicylic acid
- Jasmonic acid
- Ethylene

There can be substantial cross-talk among these pathways.

Regulation by Degradation

As with many signal transduction pathways, plant gene expression during immune responses can be regulated by degradation. This often occurs when hormone binding to hormone receptors stimulates ubiquitin-associated degradation of repressor proteins that block expression of certain genes. The net result is hormone-activated gene expression. Examples:

- Auxin: binds to receptors that then recruit and degrade repressors of transcriptional activators that stimulate auxin-specific gene expression.
- Jasmonic acid: similar to auxin, except with jasmonate receptors impacting jasmonate-response signaling mediators such as JAZ proteins.
- Gibberellic acid: Gibberellin causes receptor conformational changes and binding and degradation of Della proteins.
- Ethylene: Inhibitory phosphorylation of the EIN2 ethylene response activator is blocked by ethylene binding. When this phosphorylation is reduced, EIN2 protein is cleaved and a portion of the protein moves to the nucleus to activate ethylene-response gene expression.

Ubiquitin and E3 Signaling

Ubiquitination plays a central role in cell signaling that regulates processes including protein degradation and immunological response. Although one of the main functions of ubiquitin is to target proteins for destruction, it is also useful in signaling pathways, hormone release, apoptosis and translocation of materials throughout the cell. Ubiquitination is a component of several immune responses. Without ubiquitin's proper functioning, the invasion of pathogens and other harmful molecules would increase dramatically due to weakened immune defenses.

E3 Signaling

The E3 Ubiquitin ligase enzyme is a main component that provides specificity in protein degradation pathways, including immune signaling pathways. The E3 enzyme components can be grouped by which domains they contain and include several types. These include the Ring and U-box single subunit, HECT, and CRLs. Plant signaling pathways including immune responses are controlled by several feedback pathways, which often include negative feedback; and they can be regulated by De-ubiquitination enzymes, degradation of transcription factors and the degradation of negative regulators of transcription.

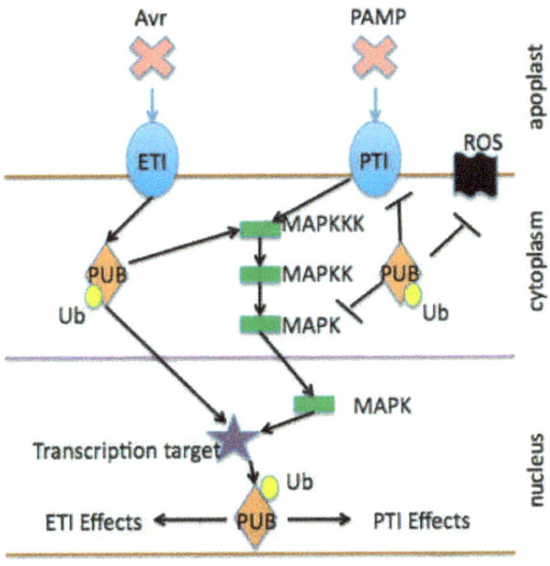

This image depicts the pathways taken during responses in plant immunity. It highlights the role and effect ubiquitin has in regulating the pathway.

Plant Breeding for Disease Resistance

Plant breeders emphasize selection and development of disease-resistant plant lines. Plant diseases can also be partially controlled by use of pesticides and by cultivation practices such as crop rotation, tillage, planting density, disease-free seeds and cleaning of equipment, but plant varieties with inherent (genetically determined) disease resistance are generally preferred. Breeding for disease resistance began when plants were first domesticated. Breeding efforts continue because pathogen populations are under selection pressure for increased virulence, new pathogens appear, evolving cultivation practices and changing climate can reduce resistance and/or strengthen pathogens, and plant breeding for other traits can disrupt prior resistance. A plant line with acceptable resistance against one pathogen may lack resistance against others.

Breeding for resistance typically includes:

- Identification of plants that may be less desirable in other ways, but which carry a useful disease resistance trait, including wild strains that often express enhanced resistance.
- Crossing of a desirable but disease-susceptible variety to another variety that is a source of resistance.
- Growth of breeding candidates in a disease-conducive setting, possibly including pathogen inoculation. Attention must be paid to the specific pathogen isolates, to address variability within a single pathogen species.
- Selection of disease-resistant individuals that retain other desirable traits such as yield, quality and including other disease resistance traits.

Resistance is termed *durable* if it continues to be effective over multiple years of widespread use as pathogen populations evolve. "Vertical resistance" is specific to certain races or strains of a pathogen species, is often controlled by single R genes and can be less durable. Horizontal or broad-spectrum resistance against an entire pathogen species is often only incompletely effective, but more durable, and is often controlled by many genes that segregate in breeding populations.

Crops such as potato, apple, banana and sugarcane are often propagated by vegetative reproduction to preserve highly desirable plant varieties, because for these species, outcrossing seriously disrupts the preferred traits. See also asexual propagation. Vegetatively propagated crops may be among the best targets for resistance improvement by the biotechnology method of plant transformation to manage genes that affect disease resistance.

Scientific breeding for disease resistance originated with Sir Rowland Biffen, who identified a single recessive gene for resistance to wheat yellow rust. Nearly every crop was then bred to include disease resistance (R) genes, many by introgression from compatible wild relatives.

GM or Transgenic Engineered Disease Resistance

The term GM ("genetically modified") is often used as a synonym of transgenic to refer to plants modified using recombinant DNA technologies. Plants with transgenic/GM disease resistance against insect pests have been extremely successful as commercial products, especially in maize and cotton, and are planted annually on over 20 million hectares in over 20 countries worldwide (see also genetically modified crops). Transgenic plant disease resistance against microbial pathogens was first demonstrated in 1986. Expression of viral coat protein gene sequences conferred virus resistance via small RNAs. This proved to be a widely applicable mechanism for inhibiting viral replication. Combining coat protein genes from three different viruses, scientists developed squash hybrids with field-validated, multiviral resistance. Similar levels of resistance to this variety of viruses had not been achieved by conventional breeding.

A similar strategy was deployed to combat papaya ringspot virus, which by 1994 threatened to destroy Hawaii's papaya industry. Field trials demonstrated excellent efficacy and high fruit quality. By 1998 the first transgenic virus-resistant papaya was approved for sale. Disease resistance has been durable for over 15 years. Transgenic papaya accounts for ~85% of Hawaiian production. The fruit is approved for sale in the U.S., Canada and Japan.

Potato lines expressing viral replicase sequences that confer resistance to potato leafroll virus were sold under the trade names NewLeaf Y and NewLeaf Plus, and were widely accepted in commercial production in 1999-2001, until McDonald's Corp. decided not to purchase GM potatoes and Monsanto decided to close their NatureMark potato business. NewLeaf Y and NewLeaf Plus potatoes carried two GM traits, as they also expressed Bt-mediated resistance to Colorado potato beetle.

No other crop with engineered disease resistance against microbial pathogens had reached the market by 2013, although more than a dozen were in some state of development and testing.

Examples of transgenic disease resistance projects

Publication year	Crop	Disease resistance	Mechanism	Development status
2012	Tomato	Bacterial spot	R gene from pepper	8 years of field trials
2012	Rice	Bacterial blight and bacterial streak	Engineered E gene	Laboratory
2012	Wheat	Powdery mildew	Overexpressed R gene from wheat	2 years of field trials at time of publication
2011	Apple	Apple scab fungus	Thionin gene from barley	4 years of field trials at time of publication
2011	Potato	Potato virus Y	Pathogen-derived resistance	1 year of field trial at time of publication
2010	Apple	Fire blight	Antibacterial protein from moth	12 years of field trials at time of publication
2010	Tomato	Multibacterial resistance	PRR from *Arabidopsis*	Laboratory scale
2010	Banana	Xanthomonas wilt	Novel gene from pepper	Now in field trial
2009	Potato	Late blight	R genes from wild relatives	3 years of field trials
2009	Potato	Late blight	R gene from wild relative	2 years of field trials at time of publication
2008	Potato	Late blight	R gene from wild relative	2 years of field trials at time of publication
2008	Plum	Plum pox virus	Pathogen-derived resistance	Regulatory approvals, no commercial sales
2005	Rice	Bacterial streak	R gene from maize	Laboratory

2002	Barley	Stem rust	Resting lymphocyte kinase (RLK) gene from resistant barley cultivar	Laboratory
1997	Papaya	Ring spot virus	Pathogen-derived resistance	Approved and commercially sold since 1998, sold into Japan since 2012
1995	Squash	Three mosaic viruses	Pathogen-derived resistance	Approved and commercially sold since 1994
1993	Potato	Potato virus X	Mammalian interferon-induced enzyme	3 years of field trials at time of publication

PRR Transfer

Research aimed at engineered resistance follows multiple strategies. One is to transfer useful PRRs into species that lack them. Identification of functional PRRs and their transfer to a recipient species that lacks an orthologous receptor could provide a general pathway to additional broadened PRR repertoires. For example, the *Arabidopsis* PRR *EF-Tu* receptor (EFR) recognizes the bacterial translation elongation factor *EF-Tu*. Research performed at Sainsbury Laboratory demonstrated that deployment of EFR into either *Nicotiana benthamiana*or *Solanum lycopersicum* (tomato), which cannot recognize *EF-Tu*, conferred resistance to a wide range of bacterial pathogens. EFR expression in tomato was especially effective against the widespread and devastating soil bacterium Ralstonia solanacearum. Conversely, the tomato PRR *Verticillium 1* (*Ve1*) gene can be transferred from tomato to *Arabidopsis*, where it confers resistance to race 1 Verticillium isolates.

Stacking

The second strategy attempts to deploy multiple NLR genes simultaneously, a breeding strategy known as stacking. Cultivars generated by either DNA-assisted molecular breeding or gene transfer will likely display more durable resistance, because pathogens would have to mutate multiple effector genes. DNA sequencing allows researchers to functionally "mine" NLR genes from multiple species/strains.

The *avrBs2* effector gene from *Xanthomona perforans* is the causal agent of bacterial spot disease of pepper and tomato. The first "effector-rationalized" search for a potentially durable R gene followed the finding that *avrBs2* is found in most disease-causing *Xanthomonas* species and is required for pathogen fitness. The *Bs2* NLR gene from the wild pepper, *Capsicum chacoense*, was moved into tomato, where it inhibited pathogen growth. Field trials demonstrated robust resistance without bactericidal chemicals. However, rare strains of *Xanthomonas* overcame *Bs2*-mediated resistance in pepper by acquisition of *avrBs2* mutations that avoid recognition but retain virulence. Stacking R genes that each recognize a different core effector could delay or prevent adaptation.

More than 50 loci in wheat strains confer disease resistance against wheat stem, leaf and yellow stripe rust pathogens. The Stem rust 35 (*Sr35*) NLR gene, cloned from a diploid relative of cultivated wheat, *Triticum monococcum*, provides resistance to wheat rust isolate *Ug99*. Similarly, *Sr33*, from the wheat relative *Aegilops tauschii*, encodes a wheat ortholog to barley *Mla* powdery mildew–resistance genes. Both genes are unusual in wheat and its relatives. Combined with the *Sr2* gene that acts additively with at least Sr33, they could provide durable disease resistance to *Ug99* and its derivatives.

Executor Genes

Another class of plant disease resistance genes opens a "trap door" that quickly kills invaded cells, stopping pathogen proliferation. Xanthomonas and Ralstonia transcription activator–like (TAL) effectors are DNA-binding proteins that activate host gene expression to enhance pathogen virulence. Both the rice and pepper lineages independently evolved TAL-effector binding sites that instead act as an executioner that induces hypersensitive host cell death when up-regulated. *Xa27* from rice and Bs3 and Bs4c from pepper, are such "executor" (or "executioner") genes that encode non-homologous plant proteins of unknown function. Executor genes are expressed only in the presence of a specific TAL effector.

Engineered executor genes were demonstrated by successfully redesigning the pepper *Bs3* promoter to contain two additional binding sites for TAL effectors from disparate pathogen strains. Subsequently, an engineered executor gene was deployed in rice by adding five TAL effector binding sites to the *Xa27* promoter. The synthetic *Xa27* construct conferred resistance against Xanthomonas bacterial blight and bacterial leaf streak species.

Host Susceptibility Alleles

Most plant pathogens reprogram host gene expression patterns to directly benefit the pathogen. Reprogrammed genes required for pathogen survival and proliferation can be thought of as "disease-susceptibility genes." Recessive resistance genes are disease-susceptibility candidates. For example, a mutation disabled an *Arabidopsis* gene encoding pectate lyase (involved in cell wall degradation), conferring resistance to the powdery mildew pathogen *Golovinomyces cichoracearum*. Similarly, the Barley *MLO* gene and spontaneously mutated pea and tomato *MLO* orthologs also confer powdery mildew resistance.

Lr34 is a gene that provides partial resistance to leaf and yellow rusts and powdery mildew in wheat. *Lr34* encodes an adenosine triphosphate (ATP)–binding cassette (ABC) transporter. The dominant allele that provides disease resistance was recently found in cultivated wheat (not in wild strains) and, like *MLO* provides broad-spectrum resistance in barley.

Natural alleles of host translation elongation initiation factors *eif4e* and *eif4g* are also recessive viral-resistance genes. Some have been deployed to control potyviruses in barley, rice, tomato, pepper, pea, lettuce and melon. The discovery prompted a successful mutant screen for chemically induced *eif4e* alleles in tomato.

Natural promoter variation can lead to the evolution of recessive disease-resistance alleles. For example, the recessive resistance gene *xa13* in rice is an allele of *Os-8N3*. *Os-8N3* is transcriptionally activated by*Xanthomonas oryzae pv. oryzae* strains that express the TAL effector *PthXo1*. The *xa13* gene has a mutated effector-binding element in its promoter that eliminates *PthXo1* binding and renders these lines resistant to strains that rely on *PthXo1*. This finding also demonstrated that *Os-8N3* is required for susceptibility.

Xa13/Os-8N3 is required for pollen development, showing that such mutant alleles can be problematic should the disease-susceptibility phenotype alter function in other processes. However, mutations in the *Os11N3* (OsSWEET14) TAL effector–binding element were made by fusing TAL effectors to nucleases (TALENs). Genome-edited rice plants with altered *Os11N3* binding sites remained resistant to *Xanthomonas oryzae pv. oryzae*, but still provided normal development function.

Gene Silencing

RNA silencing-based resistance is a powerful tool for engineering resistant crops. The advantage of RNAi as a novel gene therapy against fungal, viral and bacterial infection in plants lies in the fact that it regulates gene expression via messenger RNA degradation, translation repression and chromatin remodelling through small non-coding RNAs. Mechanistically, the silencing processes are guided by processing products of the double-stranded RNA (dsRNA) trigger, which are known as small interfering RNAs and microRNAs.

Host Range

Among the thousands of species of plant pathogenic microorganisms, only a small minority have the capacity to infect a broad range of plant species. Most pathogens instead exhibit a high degree of host-specificity. Non-host plant species are often said to express *non-host resistance*. The term *host resistance* is used when a pathogen species can be pathogenic on the host species but certain strains of that plant species resist certain strains of the pathogen species. The causes of host resistance and non-host resistance can overlap. Pathogen host range is determined, among other things, by the presence of appropriate effectors that allow colonization of a particular host. Pathogen host range can change quite suddenly if, for example, the pathogen's capacity to synthesize a host-specific toxin or effector is gained by gene shuffling/mutation, or by horizontal gene transfer.

Epidemics and Population Biology

Native populations are often characterized by substantial genotype diversity and dispersed populations (growth in a mixture with many other plant species). They also have undergone of plant-pathogen coevolution. Hence as long as novel pathogens are not introduced/do not evolve, such populations generally exhibit only a low incidence of severe disease epidemics.

Monocrop agricultural systems provide an ideal environment for pathogen evolution, because they offer a high density of target specimens with similar/identical genotypes. The rise in mobility stemming from modern transportation systems provides pathogens with access to more potential targets. Climate change can alter the viable geographic range of pathogen species and cause some diseases to become a problem in areas where the disease was previously less important.

These factors make modern agriculture more prone to disease epidemics. Common solutions include constant breeding for disease resistance, use of pesticides, use of border inspections and plant import restrictions, maintenance of significant genetic diversity within the crop gene pool (see crop diversity), and constant surveillance to accelerate initiation of appropriate responses. Some pathogen species have much greater capacity to overcome plant disease resistance than others, often because of their ability to evolve rapidly and to disperse broadly.

Chemical Defenses

Persimmon, genus *Diospyros*, has a high tannin content which gives immature fruit, seen above, an astringent and bitterflavor.

The evolution of chemical defences in plants is linked to the emergence of chemical substances that are not involved in the essential photosynthetic and metabolic activities. These substances, secondary metabolites, are organic compounds that are not directly involved in the normal growth, development or reproduction of organisms, and often produced as by-products during the synthesis of primary metabolic products. Although these secondary metabolites have been thought to play a major role in defenses against herbivores, a meta-analysis of recent relevant studies has suggested that they have either a more minimal (when compared to other non-secondary metabolites, such as primary chemistry and physiology) or more complex involvement in defense.

Secondary metabolites are often characterized as either *qualitative* or *quantitative*. Qualitative metabolites are defined as toxins that interfere with an herbivore's metabolism, often by blocking specific biochemical reactions. Qualitative chemicals are present in plants in relatively low concentrations (often less than 2% dry weight), and are not dosage dependent. They are usually small, water-soluble molecules, and therefore can be rapidly synthesized, transported and stored with relatively little energy cost to the plant. Qualitative allelochemicals are usually effective against non-adapted specialists and generalist herbivores.

Quantitative chemicals are those that are present in high concentration in plants (5 – 40% dry weight) and are equally effective against all specialists and generalist herbivores. Most quantitative metabolites are digestibility reducers that make plant cell walls indigestible to animals. The effects of quantitative metabolites are dosage dependent and the higher these chemicals' proportion in the herbivore's diet, the less nutrition the herbivore can gain from ingesting plant tissues. Because they are typically large molecules, these defenses are energetically expensive to produce and maintain, and often take longer to synthesize and transport.

The geranium, for example, produces a unique chemical compound in its petals to defend itself from Japanese beetles. Within 30 minutes of ingestion the chemical paralyzes the herbivore. While the chemical usually wears off within a few hours, during this time the beetle is often consumed by its own predators.

Types of Chemical Defences

Plants have evolved many secondary metabolites involved in plant defense, which are collectively known as antiherbivory compounds and can be classified into three sub-groups: nitrogen compounds (including *alkaloids*, *cyanogenic glycosides*, *glucosinolates* and *benzoxazinoids*), terpenoids, and phenolics.

Alkaloids are derived from various amino acids. Over 3000 known alkaloids exist, examples include nicotine, caffeine, morphine, cocaine, colchicine, ergolines, strychnine, and quinine. Alkaloids have pharmacological effects on humans and other animals. Some alkaloids can inhibit or activate enzymes, or alter carbohydrate and fat storage by inhibiting the formation phosphodiester bonds involved in their breakdown.Certain alkaloids bind to nucleic acids and can inhibit synthesis of proteins and affect DNA repair mechanisms. Alkaloids can also affect cell membrane and cytoskeletal structure causing the cells to weaken, collapse, or leak, and can affect nerve transmission. Although alkaloids act on a diversity of metabolic systems in humans and other animals, they almost uniformly invoke an aversively bitter taste.

Cyanogenic glycosides are stored in inactive forms in plant vacuoles. They become toxic when herbivores eat the plant and break cell membranes allowing the glycosides to come into contact with enzymes in the cytoplasm releasing hydrogen cyanide which blocks cellular respiration.Glucosinolates are activated in much the same way as cyanogenic glucosides, and the products can cause gastroenteritis, salivation, diarrhea, and irritation of the mouth. Benzoxazinoids, secondary defence metabolites, which are characteristic for grasses (Poaceae), are also stored as inactive glucosides in the plant vacuole. Upon tissue disruption they get into contact with β-glucosidases from the chloroplasts, which enzymatically release the toxic aglucones. Whereas some benzoxazinoids are constitutively present, others are only synthesised following herbivore infestation, and thus, considered inducible plant defenses against herbivory. The terpenoids, sometimes referred to as isoprenoids, are organic chemicals similar to terpenes, derived from five-carbon isoprene units. There are over 10,000 known types of terpenoids. Most are multicyclic structures which differ from one another in both functional groups, and in basic carbon skeletons. Monoterpenoids, continuing 2 isoprene units, are volatile essential oils such as citronella, limonene, menthol, camphor, and pinene. Diterpenoids, 4 isoprene units, are widely distributed in latex and resins, and can be quite toxic. Diterpenes are responsible for making *Rhododendron* leaves poisonous. Plant steroids and sterols are also produced from terpenoid precursors, including vitamin D, glycosides (such as digitalis) and saponins (which lyse red blood cells of herbivores).

Phenolics, sometimes called *phenols*, consist of an aromatic 6-carbon ring bonded to a hydroxy group. Some phenols have antiseptic properties, while others disrupt endocrine activity. Phenolics range from simple tannins to the more complex flavonoids that give plants much of their red, blue, yellow, and white pigments. Complex phenolics called polyphenols are capable of producing many different types of effects on humans, including antioxidant properties. Some examples of phenolics used for defense in plants are: lignin, silymarin and cannabinoids. Condensed tannins, polymers composed of 2 to 50 (or more) flavonoid molecules, inhibit herbivore digestion by binding to consumed plant proteins and making them more difficult for animals to digest, and by interfering with protein absorption and digestive enzymes. Silica and lignins, which are completely indigestible to animals, grind down insect mandibles (appendages necessary for feeding).

In addition to the three larger groups of substances mentioned above, fatty acid derivates, amino acids and even peptides are also used as defense. The cholinergic toxine, cicutoxin of water hemlock, is a polyyne derived from the fatty acid metabolism. β-N-Oxalyl-L-α,β-diaminopropionic acid as simple amino acid is used by the sweet pea which leads also to intoxication in humans. The synthesis of fluoroacetate in several plants is an example of the use of small molecules to disrupt the metabolism of herbivores, in this case the citric acid cycle.

In tropical *Sargassum* and *Turbinaria* species that are often preferentially consumed by herbivorous fishes and echinoids, there is a relatively low level of phenolics and tannins.

Mechanical Defenses

Plants have many external structural defenses that discourage herbivory. Depending on the herbivore's physical characteristics (i.e. size and defensive armor), plant structural defenses on stems and leaves can deter, injure, or kill the grazer. Some defensive compounds are produced internally but are released onto the plant's surface; for example, resins, lignins, silica, and wax cover the epidermis of terrestrial plants and alter the texture of the plant tissue. The leaves of holly plants, for instance, are very smooth and slippery making feeding difficult. Some plants produce gummosis or sap that traps insects.

The thorns on the stem of this raspberry plant, serve as a mechanical defense against herbivory.

A plant's leaves and stem may be covered with sharp prickles, spines, thorns, or trichomes- hairs on the leaf often with barbs, sometimes containing irritants or poisons. Plant structural features like spines and thorns reduce feeding by large ungulate herbivores (e.g. kudu, impala, and goats) by restricting the herbivores' feeding rate, or by wearing down the molars. Raphides are sharp needles of calcium oxalate or calcium carbonate in plant tissues, making ingestion painful, damaging a herbivore's mouth and gullet and causing more efficient delivery of the plant's toxins. The structure of a plant, its branching and leaf arrangement may also be evolved to reduce herbivore impact. The shrubs of New Zealand have evolved special wide branching adaptations believed to be a response to browsing birds such as the moas. Similarly, African Acacias have long spines low in the canopy, but very short spines high in the canopy, which is comparatively safe from herbivores such as giraffes.

Trees such as coconut and other palms, may protect their fruit by multiple layers of armor, needing efficient tools to break through to the seed contents, and special skills to climb the tall and relatively smooth trunk.

Coconut palms protect their fruit by surrounding it with multiple layers of armor.

Thigmonasty

Thigmonastic movements, those that occur in response to touch, are used as a defense in some plants. The leaves of the sensitive plant, *Mimosa pudica*, close up rapidly in response to direct touch, vibration, or even electrical and thermal stimuli. The proximate cause of this mechanical response is an abrupt change in the turgor pressure in the pulvini at the base of leaves resulting from osmotic phenomena. This is then spread via both electrical and chemical means through the plant; only a single leaflet need be disturbed.

This response lowers the surface area available to herbivores, which are presented with the underside of each leaflet, and results in a wilted appearance. It may also physically dislodge small herbivores, such as insects.

Mimicry and camouflage

Some plants mimic the presence of insect eggs on their leaves, dissuading insect species from laying their eggs there. Because female butterflies are less likely to lay their eggs on plants that already have butterfly eggs, some species of neotropical vines of the genus *Passiflora* (Passion flowers) contain physical structures resembling the yellow eggs of *Heliconius* butterflies on their leaves, which discourage oviposition by butterflies.

Indirect Defenses

The large thorn-like stipules of *Acacia collinsii* are hollow and offer shelter for ants, which in return protect the plant against herbivores.

Another category of plant defenses are those features that indirectly protect the plant by enhancing the probability of attracting the natural enemies of herbivores. Such an arrangement is known as mutualism, in this case of the "enemy of my enemy" variety. One such feature are semiochemicals, given off by plants. Semiochemicals are a group of volatile organic compounds involved in interactions between organisms. One group of semiochemicals are allelochemicals; consisting of allomones, which play a defensive role in interspecies communication, and kairomones, which are used by members of higher trophic levels to locate food sources. When a plant is attacked it releases allelochemics containing an abnormal ratio of these herbivore-induced plant volatiles (HIPVs). Predators sense these volatiles as food cues, attracting them to the damaged plant, and to feeding herbivores. The subsequent reduction in the number of herbivores confers a fitness benefit to the plant and demonstrates the indirect defensive capabilities of semiochemicals. Induced volatiles also have drawbacks, however; some studies have suggested that these volatiles also attract herbivores.

Plants also provide housing and food items for natural enemies of herbivores, known as "biotic" defense mechanisms, as a means to maintain their presence. For example, trees from the genus *Macaranga* have adapted their thin stem walls to create ideal housing for an ant species (genus *Crematogaster*), which, in turn, protects the plant from herbivores.

In addition to providing housing, the plant also provides the ant with its exclusive food source; from the food bodies produced by the plant. Similarly, some *Acacia* tree species have developed thorns that are swollen at the base, forming a hollowing structure that acts as housing. These *Acacia* trees also produce nectar in extrafloral nectaries on their leaves as food for the ants.

Plant use of endophytic fungi in defense is a very common phenomenon. Most plants have endophytes, microbial organisms that live within them. While some cause disease, others protect plants from herbivores and pathogenic microbes. Endophytes can help the plant by producing toxins harmful to other organisms that would attack the plant, such as alkaloid producing fungi which are common in grasses such as tall fescue (*Festuca arundinacea*).

Leaf Shedding and Color

There have been suggestions that leaf shedding may be a response that provides protection against diseases and certain kinds of pests such as leaf miners and gall forming insects. Other responses such as the change of leaf colors prior to fall have also been suggested as adaptations that may help undermine the camouflage of herbivores. Autumn leaf color has also been suggested to act as an honest warning signal of defensive commitment towards insect pests that migrate to the trees in autumn.

Costs and benefits

Defensive structures and chemicals are costly as they require resources that could otherwise be used by plants to maximize growth and reproduction. Many models have been proposed to explore how and why some plants make this investment in defenses against herbivores.

Optimal Defense Hypothesis

The optimal defense hypothesis attempts to explain how the kinds of defenses a particular plant might use reflect the threats each individual plant faces. This model considers three main factors, namely: risk of attack, value of the plant part, and the cost of defense.

The first factor determining optimal defense is risk: how likely is it that a plant or certain plant parts will be attacked? This is also related to the *plant apparency hypothesis*, which states that a plant will invest heavily in broadly effective defenses when the plant is easily found by herbivores.Examples of apparent plants that produce generalized protections include long-living trees, shrubs, and perennial grasses. Unapparent plants, such as short-lived plants of early successional stages, on the other hand, preferentially invest in small amounts of qualitative toxins that are effective against all but the most specialized herbivores.

The second factor is the value of protection: would the plant be less able to survive and reproduce after removal of part of its structure by a herbivore? Not all plant parts are of equal evolutionary value, thus valuable parts contain more defenses. A plant's stage of development at the time of feeding also affects the resulting change in fitness.

Experimentally, the fitness value of a plant structure is determined by removing that part of the plant and observing the effect. In general, reproductive parts are not as easily replaced as vegetative parts, terminal leaves have greater value than basal leaves, and the loss of plant parts mid-season has a greater negative effect on fitness than removal at the beginning or end of the season. Seeds in particular tend to be very well protected. For example, the seeds of many edible fruits and nuts contain cyanogenic glycosides such as amygdalin. This results from the need to balance the effort needed to make the fruit attractive to animal dispersers while ensuring that the seeds are not destroyed by the animal.

The final consideration is cost: how much will a particular defensive strategy cost a plant in energy and materials? This is particularly important, as energy spent on defense cannot be used for other functions, such as reproduction and growth. The optimal defense hypothesis predicts that plants will allocate more energy towards defense when the benefits of protection outweigh the costs, specifically in situations where there is high herbivore pressure.

Carbon:Nutrient Balance Hypothesis

The carbon:nutrient balance hypothesis, also known as the *environmental constraint hypothesis* or *Carbon Nutrient Balance Model* (CNBM), states that the various types of plant defenses are responses to variations in the levels of nutrients in the environment. This hypothesis predicts the Carbon/Nitrogen ratio in plants determines which secondary metabolites will be synthesized. For example, plants growing in nitrogen-poor soils will use carbon-based defenses (mostly digestibility reducers), while those growing in low-carbon environments (such as shady conditions) are more likely to produce nitrogen-based toxins. The hypothesis further predicts that plants can change their defenses in response to changes in nutrients. For example, if plants are grown in low-nitrogen conditions, then these plants will implement a defensive strategy composed of constitutive carbon-based defenses. If nutrient levels subsequently increase, by for example the addition of fertilizers, these carbon-based defenses will decrease.

Growth Rate Hypothesis

The growth rate hypothesis, also known as the *resource availability hypothesis*, states that defense strategies are determined by the inherent growth rate of the plant, which is in turn determined by the resources available to the plant. A major assumption is that available resources are the limiting factor in determining the maximum growth rate of a plant species. This model predicts that the level of defense investment will increase as the potential of growth decreases. Additionally, plants in resource-poor areas, with inherently slow-growth rates, tend to have long-lived leaves and twigs, and the loss of plant appendages may result in a loss of scarce and valuable nutrients.

A recent test of this model involved a reciprocal transplants of seedlings of 20 species of trees between clay soils (nutrient rich) and white sand (nutrient poor) to determine whether trade-offs between growth rate and defenses restrict species to one habitat. When planted in white sand and protected from herbivores, seedlings originating from clay outgrew those originating from the nutrient-poor sand, but in the presence of herbivores the seedlings originating from white sand performed better, likely due to their higher levels of constitutive carbon-based defenses. These finding suggest that defensive strategies limit the habitats of some plants.

Growth-Differentiation Balance Hypothesis

The growth-differentiation balance hypothesis states that plant defenses are a result of a tradeoff between "growth-related processes" and "differentiation-related processes" in different environments. Differentiation-related processes are defined as "processes that enhance the structure or function of existing cells (i.e. maturation and specialization)." A plant will produce chemical defenses only when energy is available from photosynthesis, and plants with the highest concentrations of secondary metabolites are the ones with an intermediate level of available resources. The GDBH also accounts for tradeoffs between growth and defense over a resource availability gradient. In situations where resources (e.g. water and nutrients) limit photosynthesis, carbon supply is predicted to limit both growth and defense. As resource availability increases, the requirements needed to support photosynthesis are met, allowing for accumulation of carbohydrate in tissues. As resources are not sufficient to meet the large demands of growth, these carbon compounds can instead be partitioned into the synthesis of carbon based secondary metabolites (phenolics, tannins, etc.). In environments where the resource demands for growth are met, carbon is allocated to rapidly dividing meristems (high sink strength) at the expense of secondary metabolism. Thus rapidly growing plants are predicted to contain lower levels of secondary metabolites and vice versa. In addition, the tradeoff predicted by the GDBH may change over time, as evidenced by a recent study on Salix spp. Much support for this hypothesis is present in the literature, and some scientists consider the GDBH the most mature of the plant defense hypotheses.

Importance to Humans

Agriculture

The variation of plant susceptibility to pests was probably known even in the early stages of agriculture in humans. In historic times, the observation of such variations in susceptibility have provided solutions for major socio-economic problems. The grape phylloxera was introduced from North America to France in 1860 and in 25 years it destroyed nearly a third (100,000 km^2) of the French grape yards. Charles Valentine Riley noted that the American species *Vitis labrusca* was resistant to *Phylloxera*. Riley, with J. E. Planchon, helped save the French wine industry by suggesting the grafting of the susceptible but high quality grapes onto *Vitis labrusca* root stocks. The formal study of plant resistance to herbivory was first covered extensively in 1951 by Reginald (R.H.) Painter, who is widely regarded as the founder of this area of research, in his book *Plant Resistance to Insects*. While this work pioneered further research in the US, the work of Chesnokov was the basis of further research in the USSR.

Fresh growth of grass is sometimes high in prussic acid content and can cause poisoning of grazing livestock. The production of cyanogenic chemicals in grasses is primarily a defense against herbivores.

The human innovation of cooking may have been particularly helpful in overcoming many of the defensive chemicals of plants. Many enzyme inhibitors in cereal grains and pulses, such as trypsin inhibitors prevalent in pulse crops, are denatured by cooking, making them digestible.

It has been known since the late 17th century that plants contain noxious chemicals which are avoided by insects. These chemicals have been used by man as early insecticides; in 1690 nicotine was extracted from tobacco and used as a contact insecticide. In 1773, insect infested plants were treated with nicotine fumigation by heating tobacco and blowing the smoke over the plants. The flowers of *Chrysanthemum* species contain pyrethrin which is a potent insecticide. In later years, the applications of plant resistance became an important area of research in agriculture and plant breeding, particularly because they can serve as a safe and low-cost alternative to the use of pesticides. The important role of secondary plant substances in plant defense was described in the late 1950s by Vincent Dethier and G.S. Fraenkel. The use of botanical pesticides is widespread and notable examples include Azadirachtin from the neem (*Azadirachta indica*), d-Limonene from Citrus species, Rotenone from *Derris*, Capsaicin from chili pepper and Pyrethrum.

Natural materials found in the environment also induce plant resistance as well. Chitosan derived from chitin induce a plant's natural defense response against pathogens, diseases and insects including cyst nematodes, both are approved as biopesticides by the EPA to reduce the dependence on toxic pesticides.

The selective breeding of crop plants often involves selection against the plant's intrinsic resistance strategies. This makes crop plant varieties particularly susceptible to pests unlike their wild relatives. In breeding for host-plant resistance, it is often the wild relatives that provide the source of resistance genes. These genes are incorporated using conventional approaches to plant breeding, but have also been augmented by recombinant techniques, which allow introduction of genes from completely unrelated organisms. The most famous transgenic approach is the introduction of genes from the bacterial species, *Bacillus thuringiensis*, into plants. The bacterium produces proteins that, when ingested, kill lepidopteran caterpillars. The gene encoding for these highly toxic proteins, when introduced into the host plant genome, confers resistance against caterpillars, when the same toxic proteins are produced within the plant. This approach is controversial, however, due to the possibility of ecological and toxicological side effects.

Pharmaceutical

Many currently available pharmaceuticals are derived from the secondary metabolites plants use to protect themselves from herbivores, including opium, aspirin, cocaine, and atropine. These chemicals have evolved to affect the biochemistry of insects in very specific ways. However, many of these biochemical pathways are conserved in vertebrates, including humans, and the chemicals act on human biochemistry in ways similar to that of insects. It has therefore been suggested that the study of plant-insect interactions may help in bioprospecting.

There is evidence that humans began using plant alkaloids in medical preparations as early as 3000 B.C.Although the active components of most medicinal plants have been isolated only recently (beginning in the early 19th century) these substances have been used as drugs throughout the human history in potions, medicines, teas and as poisons. For example, to combat herbivory by the larvae of some Lepidoptera species, Cinchona trees produce a variety of alkaloids, the most familiar of which is quinine. Quinine is extremely bitter, making the bark of the tree quite unpalatable, it is also an anti-fever agent, known as Jesuit's bark, and is especially useful in treating malaria.

Illustration from the 15th-century manuscript *Tacuinum Sanitatis* detailing the beneficial and harmful properties of Mandrakes

Throughout history mandrakes (*Mandragora officinarum*) have been highly sought after for their reputed aphrodisiac properties. However, the roots of the mandrake plant also contain large quantities of the alkaloid scopolamine, which, at high doses, acts as a central nervous system depressant, and makes the plant highly toxic to herbivores. Scopolamine was later found to be medicinally used for pain management prior to and during labor; in smaller doses it is used to prevent motion sickness.One of the most well-known medicinally valuable terpenes is an anticancer drug, taxol, isolated from the bark of the Pacific yew, *Taxus brevifolia*, in the early 1960s.

Biological Pest Control

Repellent companion planting, defensive live fencing hedges, and "obstructive-repellent" interplanting, with host-plant resistance species as beneficial 'biological control agents' is a technique in biological pest control programs for: organic gardening, wildlife gardening, sustainable gardening, and sustainable landscaping; in organic farming and sustainable agriculture; and in restoration ecology methods for habitat restoration projects.

DISEASE RESISTANCE IN FRUIT AND VEGETABLES

There are a number of lines of defence against pests (that, those animals that cause damage to the plants we grow) and diseases in the O, principal among these being the practice of good husbandry, creating healthy soil and ensuring high standards of garden hygiene. But no matter how diverse and healthy the garden eco-system may be, there will always be a degree of disease and pest presence. In many ways, some level of pathogen population in the garden can be not only acceptable but desirable as they are indicative of a generally healthful and diverse environment, and add to the overall robustness of the system as an immunity to such detrimental influences will build up, particularly in a balanced polycultural regime. Indeed, most of the plants we grow will tend to be selected because they are trouble free, and those that are more susceptible to attack will have fallen by the wayside over time. However, most farmers find it unacceptable that the food crops they grow are damaged by pests.

For these crops there has been considerable research and selective breeding carried out in order to find cultivars that are resistant or immune to pest and disease damage. Breeding for plant disease resistance generally has involved finding suitable genetic material amongst existing stocks or in the wild, which is then incorporated into commercial varieties.

Example: The Apple

In the case of apples, in which research is being carried out in order to develop resistance to diseases such as apple scab (*Venturia inaequalis*), powdery mildew (*Podosphaera leucotricha*), orchard fireblight (*Erwinia amylovora*), woolly apple aphid (*Eriosoma lanigerum*) and collar rot (*Phytophthora cactorum*), the main sources of resistant material used in breeding programmes such as those being run by East Malling Research Station in England or Hortresearch in New Zealand are major gene resistances derived from crab-apples. The Vf gene for black spot resistance is derived from the ornamental crab-apple species *Malus floribunda*. Most black spot resistant cultivars developed around the world carry this gene, but there are some selections that carry the Vr (from M. pumila) or Vm (from M. micromalus) gene. Major gene resistances to powdery mildew are derived from M. robusta (Pl1) and M. zumi (Pl2), and the apple cultivar *Northern Spy* has a long-standing reputation for its major gene resistance to woolly apple aphid. Since early this century this resistance has been used to develop woolly aphid resistant rootstocks such as MM.106 and M.793. Much later it was shown that the cultivar was also very resistant to collar rot and a useful breeding parent for this resistance.

Resistance and immunity

Some plants can tolerate the presence of large numbers of insects without being severely affected. This is not very satisfactory however as insects will still cause damage, and in fact further breeding and population expansion of the pest species is supported. Other varieties are less attractive to pests, but this can be difficult to sustain or demonstrate. The most valuable form of resistance is where the pest cannot survive as well on one variety as on another. In some cases this can actually make the plants immune to attack, as is the case with the lettuces *Avoncrisp* and *Avondefiance* which were bred at the Institute of Horticultural Research, Wellesbourne during the 1960s, which are fully resistant to lettuce root aphid (*Pemphigus bursarius*).

Trade-Off of Breeding for Resistance

Sometimes however there can be a *trade-off*, for those varieties which have increased immunity or resistance may be lacking in other qualities such as flavour, yield or quality. Celery resistant to the Fusarium fungus (*Fusarium oxysporum spp.*) may not succumb to this disease, but may also be unacceptably short, ribby and low yielding. Further, a cultivar that is resistant to one disease may be more susceptible to another that is equally important. A lettuce cultivar that is resistant to mosaic virus may be sensitive to corky root disease, whilst another that resists corky root may be vulnerable to downy mildew (*Brim lactic*). Another drawback to resistance is that depending on the host pathogen system, resistance is sometimes not long lasting as new pathogen strains quickly develop, and further research and breeding is constantly needed.

Availability of Resistant Varieties

Resistant varieties are not available for all crops. For several of the most damaging plant diseases, such as Potato blight (*Phytophthora infestans*) and white rot (*Sclerotic cepivorum*) of the Allium family, no acceptable resistant cultivars are yet available. In addition, commercial seed companies and plant breeders rarely invest resources into developing resistant cultivars for more minor or speciality crops, which often tend to be those of greater interest to the organic grower.

In general it is probably fair to say that resistance will not fully guarantee total crop protection, but choosing resistant varieties should rather be considered as a part of an overall Integrated pest management strategy, especially against virus diseases. In particular they can be especially useful where the threat from specific pests and diseases is high.

Table of Resistant Crops

The table below illustrates a number of pest and disease resistant vegetable crop varieties, although should not be considered comprehensive.

Crop	Variety	Comments
French Bean	Aigullion	Resistant to bean mosaic virus and anthracnose
	Hildora	Tolerant of bean mosaic virus
	Maxi	Tolerant of bean mosaic virus
	Nektar Queen	Resistant to bean mosaic virus
	Farba	Resistant to bean mosaic virus
	Hilda	Resistant to bean mosaic virus
Broad Bean	Futura	Tolerant of chocolate spot
Brussels sprouts	Cavalier	Resistant to light leaf spot
	Braveheart F1	Resistant to powdery mildew and light leaf spot
Cabbage	Stonehead F1	Resistant to mildew
Calabrese	Emperor F1	Highly tolerant of black rot and downy mildew
	Sampson F1	Highly resistant to black rot and downy mildew
Carrot	Fly Away F1	Bred for resistance to carrot fly

	Newmarket	Good resistance to splitting
Cucumber	Bush Champion F1	Resistant to cucumber mosaic virus
	Marketmore	Resistant to cucumber mosaic virus
	Burpless Tasty Green F1	Resistant to mildew and tolerant of heat
	Slice King F1	Resistant to downy and powdery mildew, gummosis and angular leaf spot
	Cumlaude F1	Tolerant of powdery mildew
	Media F1	Tolerant of powdery mildew
Leek	Swiss Giant, Zermatt	Moderate rust resistance
	Alvitta	Rust Tolerant
	Conora	Good resistance to rust
Lettuce	Alexandria	Highly resistant to downy mildew and bolting
	Lisbusa	Resistant to downy mildew, tolerant of mosaic virus, root aphid and tipburn
	Sylvestra	Resistant to downy mildew and aphids
	Little Gem	Resistant to root aphid
	Barcelona	Resistant to aphids and mildew
	Iglo	Resistant to downey mildew
	Revolution	Resistant to mildew and bolting
	Smile	Resistant to aphids, mildew, tipburn and bolting
Parsnip	Avonreister	Very good resistance to canker
	Gladiator F1	High resistance to canker
	Tender and True	Good canker resistance
Pea	Ambassador	Resistant to powdery mildew, entation virus and fusarium wilt
	Cavalier	Good resistance to powdery mildew.
	Greenshaft	Resistant to downy mildew and fusarium wilt
	Rondo	Resistant to fusarium wilt
Potato (first early)	Colleen	Good resistance to blight and scab
	Premiere	Resistant to blight and spraing
	Pentland Javelin	Good all round disease resistance, including golden eelworm and scab
	Swift	Resistant to golden eelworm and tolerant to blackleg
Potato (second early)	Cosmos	Resistant to blight and common scab
	Osprey	Very resistant to scab and eelworm
Potato (maincrop)	Harmony	Very resistant to scab and partially resistant to white and golden eelworm
	Milva	Blight resistant
	Remarka	Good all around disease resistance

	Admiral	Good resistance to blight and scab
	Cara	Blight resistant
	Valor	Good overall disease resistance combined with high eelworm and blight resistance
Pepper	Bell Boy F1	Mosaic virus resistant
	Mandy F1	Highly resistant to mosaic virus
Sweetcorn	Golden sweet F1	Highly resistant to smut
Tomato	Alicante	Good resistance to greenback and mildew
	Libra F1	Resistant to fusarium crown and root rot
	Shirley F1	Resistant to fusarium, cladosporium and TMV

REFERENCES

1. Agrios, George N. (1972). Plant Pathology (3rd ed.). Academic Press.
2. Martinelli, F., Scalenghe, R., Davino, S., Panno, S., Scuderi, G., Ruisi, P., Villa, P., Stroppiana, D., Boschetti, M., Goulart, L.R., Davis, C.E., Dandekar, A.M. (2014). "Advanced methods of plant disease detection. A review". Agronomy for Sustainable Development. 35: 1–25. doi:10.1007/s13593-014-0246-1.
3. Nicole Davis (September 9, 2009). "Genome of Irish potato famine pathogen decoded". Haas et al. Broad Institute of MIT and Harvard. Retrieved 24 July 2012.
4. Kamoun, S.; Furzer, O.; Jones, J. D. G.; Judelson, H. S.; Ali, G. S.; Dalio, R. J. D.; Roy, S. G.; Schena, L.; Zambounis, A.; Panabières, F.; Cahill, D.; Ruocco, M.; Figueiredo, A.; Chen, X. R.; Hulvey, J.; Stam, R.; Lamour, K.; Gijzen, M.; Tyler, B. M.; Grünwald, N. J.; Mukhtar, M. S.; Tomé, D. F. A.; Tör, M.; Van Den Ackerveken, G.; McDowell, J.; Daayf, F.; Fry, W. E.; Lindqvist-Kreuze, H.; Meijer, H. J. G.; et al. (2015). "The Top 10 oomycete pathogens in molecular plant pathology". Molecular Plant Pathology: n/a. doi:10.1111/mpp.12190.
5. Grünwald, N. J.; Goss, E. M.; Press, C. M. (2008). "Phytophthora ramorum: A pathogen with a remarkably wide host range causing sudden oak death on oaks and ramorum blight on woody ornamentals". Molecular Plant Pathology. 9 (6): 729. doi:10.1111/J.1364-3703.2008.00500.X.
6. "Scientists discover how deadly fungal microbes enter host cells". (VBI) at Virginia Tech affiliates. Physorg. July 22, 2010. Retrieved July 31, 2012.
7. Jackson RW (editor). (2009). Plant Pathogenic Bacteria: Genomics and Molecular Biology. Caister Academic Press. ISBN 978-1-904455-37-0.
8. Burkholder (October 1948). "Bacteria as Plant Pathogens". Annual Review of Microbiology. Cornell University. 2: 389–412. doi:10.1146/annurev.mi.02.100148.002133. PMID 18104350.
9. "Research team unravels tomato pathogen's tricks of the trade". Virginia Tech. 2011.

10. Creamer, Rebecca; H. Hubble; A. Lewis (May 2005). "Curtovirus Infection of Chile Pepper in New Mexico". Plant Diseases. 89 (5): 480–486. doi:10.1094/PD-89-0480.
11. Huynh, Bao-Lam; Matthews, William C.; Ehlers, Jeffrey D.; Lucas, Mitchell R.; Santos, Jansen R. P.; Ndeve, Arsenio; Close, Timothy J.; Roberts, Philip A. (2015-10-08). "A major QTL corresponding to the Rk locus for resistance to root-knot nematodes in cowpea (Vigna unguiculata L. Walp.)". Theoretical and Applied Genetics. 129 (1): 87–95. doi:10.1007/s00122-015-2611-0. ISSN 0040-5752. PMC 4703619 . PMID 26450274.
12. Jankevicius, J.V. et al. Ciclo biológico de Phytomonas / Biological cycle of Phytomonas. Memórias do Instituto Oswaldo Cruz, Rio de Janeiro, v. 83, supl. 1, 1988.
13. Ma, Winbo (March 28, 2011). "How do plants fight disease? Breakthrough research by UC Riverside plant pathologist offers a clue". UC Riverside.
14. "1st large-scale map of a plant's protein network addresses evolution, disease process". Dana-Farber Cancer Institute. July 29, 2011. Retrieved 24 July 2012.
15. "History of Plant Pathology". Retrieved 5 February 2015.
16. "Plasmopara viticola, the Cause of Downy Mildew of Grapes". The Origin of Plant Pathology and The Potato Famine, and Other Stories of Plant Diseases. Retrieved 4 February 2015.
17. "Fusarium oxysporum : The End of the Banana Industry?". The Origin of Plant Pathology and The Potato Famine, and Other Stories of Plant Diseases. Retrieved 4 February 2015.
18. "Plant Diseases - History Of Plant Pathology". Retrieved 5 February 2015.
19. Aisnworth GC. (1981). Introduction to the History of Plant Pathology. Cambridge University Press. ISBN 0-521-23032-2.
20. Boyd, Jade (2012). "A bit touchy: Plants' insect defenses activated by touch". Rice University. http://news.rice.edu/2012/04/09/a-bit-touchy-plants-insect-defenses-activated-by-touch-2/
21. Venturi, S.; Donati, F.M.; Venturi, A.; Venturi, M. (2000). "Environmental Iodine Deficiency: A Challenge to the Evolution of Terrestrial Life?". Thyroid. 10 (8): 727–9. doi:10.1089/10507250050137851. PMID 11014322.
22. Venturi, Sebastiano (2011). "Evolutionary Significance of Iodine". Current Chemical Biology-. 5 (3): 155–162. doi:10.2174/187231311796765012. ISSN 1872-3136.
23. ^ Jump up to:a b c d Ehrlich, Paul R.; Peter H. Raven (December 1964). "Butterflies and plants: a study of coevolution.". Evolution. Society for the Study of Evolution. 18 (4): 586–608. doi:10.2307/2406212. JSTOR 2406212.
24. Labandeira, C.C.; D.L. Dilcher, D.R. Davis, D.L. Wagner; Davis, D. R.; Wagner, D. L. (1994). "Ninety-seven million years of angiosperm-insect association: paleobiological insights into the meaning of coevolution"(PDF). Proceedings of

the National Academy of Sciences of the United States of America. 91 (25): 12278–82. Bibcode:1994PNAS...9112278L. doi:10.1073/pnas.91.25.12278. PMC 45420 . PMID 11607501.

25. Keddy, P.A. 2007. Plants and Vegetation: Origins, Processes, Consequences. Cambridge University Press, Cambridge, UK. 666 p. Chapter 7.
26. Labandeira, C.C. (1998). "Early History Of Arthropod And Vascular Plant Associations 1". Annual Review of Earth and Planetary Sciences. 26 (1): 329–377. Bibcode:1998AREPS..26..329L. doi:10.1146/annurev.earth.26.1.329.
27. Labandeira, C. (2007). "The origin of herbivory on land: Initial patterns of plant tissue consumption by arthropods". Insect Science. 14 (4): 259–275. doi:10.1111/j.1744-7917.2007.00152.x.
28. Karban, Richard; Anurag A. Agrawal (November 2002). "Herbivore offense". Annual Review of Ecology and Systematics. 33 (1): 641–664. doi:10.1146/annurev.ecolsys.33.010802.150443.
29. Futuyma, Douglas J.; Montgomery Slatkin (1983). Coevolution. Sunderland, Massachusetts: Sinauer Associates. ISBN 0-87893-228-3.
30. Thompson, J. (1999). "What we know and do not know about coevolution: insect herbivores and plants as a test case.". In H. Olff; V. K. Brown; R. H. Drent. Herbivores: between plants and predators; the 38th symposium of the British Ecological Society in cooperation with the Netherlands Ecological Society held at the Wageningen Agricultural University, The Netherlands, 1997. Oxford: Blackwell Science. pp. 7–30. ISBN 0-632-05155-8.
31. Farrell, Brian D.; Charles Mitter (1994). "Adaptive Radiation in Insects and Plants: Time and Opportunity". American Zoologist. 34 (1): 57–69. doi:10.1093/icb/34.1.57.
32. Traw, Brian M.; Todd E. Dawson (May 2002). "Differential induction of trichomes by three herbivores of black mustard" (PDF). Oecologia. 131 (4): 526–532. doi:10.1007/s00442-002-0924-6. Retrieved 2007-05-27.
33. Walling, L.L. (2000). "The myriad plant responses to herbivores". J. Plant Growth Regul. 19: 195–216.
34. Wu, J.; Baldwin, I.T. (2009). "Herbivory-induced signalling in plants: Perception and action". Plant Cell Environ. 32: 1161–1174. doi:10.1111/j.1365-3040.2009.01943.x.
35. Sarmento, R.A.; Lemos, F.; Dias, C.R.; Kikuchi, W.T.; Rodrigues, J.C.P.; Pallini, A.; Sabelis, M.W.; Janssen, A. "A herbivorous mite down-regulates plant defence and produces web to exclude competitors. PLoS One 2011; 6, doi:10.1371/journal.pone.0023757
36. Sangha, J.S.; Yolanda; Chen, H.; Kaur, Jatinder; Khan, Wajahatullah; Abduljaleel, Zainularifeen; Alanazi, Mohammed S.; Mills, Aaron; Adalla, Candida B.; Bennett, John; Prithiviraj, Balakrishnan; Jahn, Gary C.; Leung, Hei (2013). "Proteome Analysis of Rice (Oryza sativa L.) Mutants Reveals Differentially Induced Proteins during Brown Planthopper (Nilaparvata lugens) Infestation". Int. J. Mol. Sci. 14: 3921–3945. doi:10.3390/ijms14023921. PMC 3588078 . PMID 23434671.

37. Karban, Richard; Anurag A. Agrawal; Marc Mangel (July 1997). "The benefits of induced defenses against herbivores" (PDF). Ecology. Ecological Society of America. 78 (5): 1351–1355. doi:10.2307/2266130. JSTOR 2266130. Retrieved 2007-05-27.
38. Fraenkel, G. (1959). "The raison d'être of secondary plant substances". Science. 129 (3361): 1466–70. Bibcode:1959Sci...129.1466F. doi:10.1126/science.129.3361.1466. PMID 13658975.
39. Whittaker, Robert H. (1970). "The biochemical ecology of higher plants". In Ernest Sondheimer; John B. Simeone. Chemical ecology. Boston: Academic Press. pp. 43–70. ISBN 0-12-654750-5.
40. Whittaker, Robert H. (1975). Communities and ecosystems. New York: Macmillan. ISBN 0-02-427390-2.
41. Carmona, Diego; Marc J. Lajeunesse; Marc T.J. Johnson (April 2011). "Plant traits that predict resistance to herbivores" (PDF). Functional Ecology. 25 (2): 358–367. doi:10.1111/j.1365-2435.2010.01794.x. Retrieved 26 June 2011.
42. Theis, Nina; Manuel Lerdau (2003). "The evolution of function in plant secondary metabolites" (PDF). International Journal of Plant Science. 164 (3 Suppl.): S93–S102. doi:10.1086/374190. Archived from the original (PDF) on 2007-04-18. Retrieved 2007-05-27.
43. "Geraniums and Begonias". Retrieved 2014-01-15.
44. "Biochemical defenses: secondary metabolites". Plant Defense Systems & Medicinal Botany. Retrieved 2007-05-21.
45. "Alkaloids: contain a N-containing heterocycle". Plant Defense Systems & Medicinal Botany. Retrieved 2007-06-26.
46. Roberts, Margaret F.; Michael Wink (1998). Alkaloids: biochemistry, ecology, and medicinal applications. New York: Plenum Press. ISBN 0-306-45465-3.
47. Sneden, Albert T. "Alkaloids". Natural Products as Medicinally Useful Agents. Retrieved 2007-05-21.
48. Rhoades, David F (1979). "Evolution of Plant Chemical Defense against Herbivores". In Rosenthal, Gerald A.; Janzen, Daniel H. Herbivores: Their Interaction with Secondary Plant Metabolites. New York: Academic Press. pp. 3–54. ISBN 0-12-597180-X.
49. Toxicon Volume 38, Issue 1, January 2000, Pages 11-36 János Vetter Plant cyanogenic glycosides doi:10.1016/S0041-0101(99)00128-2
50. Niemeyer, HM (2009). "Hydroxamic acids derived from 2-hydroxy-2H-1,4-benzoxazin-3(4H)-one: key defense chemicals of cereals". J Agric Food Chem. 57 (5): 1677–1696. doi:10.1021/jf8034034.
51. Glauser, G; Marti, G; Villard, N; Doyen, GA; Wolfender, J-L; Turlings, TCJ; Erb, M (2011). "Induction and detoxification of maize 1,4-benzoxazin-3-ones by insect herbivores". Plant Journal. 68 (5): 901–911. doi:10.1111/j.1365-313X.2011.04740.x.
52. "Terpenoids". Plant Defense Systems & Medicinal Botany. Retrieved 2007-06-26.

53. Gershenzon, Jonathan; Wolfgang Kreis (1999). "Biochemistry of terpenoids". In Michael Wink. Biochemistry of plant secondary metabolism. London: Sheffield Academic Press. pp. 222–279. ISBN 0-8493-4085-3.
54. Sneden, Albert T. "Terpenes". Natural Products as Medicinally Useful Agents. Retrieved 2007-05-21.
55. "Phenols". Plant Defense Systems & Medicinal Botany. Retrieved 2007-05-21.
56. Van Soest, Peter J. (1982). Nutritional ecology of the ruminant: ruminant metabolism, nutritional strategies, the cellulolytic fermentation, and the chemistry of forages and plant fibers. Corvallis, Oregon: O & B Books. ISBN 0-9601586-0-X.
57. John W. Hylin (1969). "Toxic peptides and amino acids in foods and feeds". Journal of Agricultural and Food Chemistry. 17 (3): 492–496. doi:10.1021/jf60163a003.
58. E. Anet; B. Lythgoe; M. H. Silk; S. Trippett (1953). "Oenanthotoxin and cicutoxin. Isolation and structures". Journal of the Chemical Society: 309–322. doi:10.1039/JR9530000309.
59. Mark V. Barrow; Charles F. Simpson; Edward J. Miller (1974). "Lathyrism: A Review". The Quarterly Review of Biology. 49 (2): 101–128. doi:10.1086/408017. JSTOR 2820941. PMID 4601279.
60. Donald A. Levin; King, Dennis R. (1991). "The Impact of Fluoroacetate-Bearing Vegetation on Native Australian Fauna: A Review". Oikos. Blackwell Publishing. 61 (3): 412–430. doi:10.2307/3545249. JSTOR 3545249.
61. Steinberg, Peter D. (1986). "Chemical defenses and the susceptibility of tropical marine brown algae to herbivores". Oecologia. 69 (4): 628–630. doi:10.1007/BF00410374.
62. Fernandes GW (1994). "Plant mechanical defenses against insect herbivory". Revista Brasileira de Entomologia. 38 (2): 421–433 .
63. Cooper, Susan M.; Norman Owen-Smith (September 1986). "Effects of plant spinescence on large mammalian herbivores". Oecologia. 68 (3): 446–455. doi:10.1007/BF01036753.
64. Bond W, Lee W & Craine J (2004). "Plant structural defences against browsing birds: a legacy of New Zealand's extinct moas". Oikos. 104 (3): 500–508. doi:10.1111/j.0030-1299.2004.12720.x.
65. Young, Truman P. (1987). "Increased thorn length in Acacia drepanolobium-an induced response to browsing". Oecologia. 71 (3): 436–438. doi:10.1007/BF00378718.
66. Young, Truman P.; Bell Okello (1998). "Relaxation of an induced defense after exclusion of herbivores: spines on Acacia drepanolobium". Oecologia. 115 (4): 508–513. doi:10.1007/s004420050548.
67. ^ Jump up to:a b Raven, Peter H.; Ray F. Evert; Susan E. Eichhorn (2005). Biology of Plants. New York: W. H. Freeman and Company. ISBN 0-7167-1007-2.
68. Williams, Kathy S.; Lawrence E. Gilbert (April 1981). "Insects as selective

agents on plant vegetative morphology: egg mimicry reduces egg-laying by butterflies". Science. 212 (4493): 467–469. Bibcode:1981Sci...212..467W. doi:10.1126/science.212.4493.467. PMID 17802547.

69. Dicke, Marcel; Joop J.A. van Loon (December 2000). "Multitrophic effects of herbivore-induced plant volatiles in an evolutionary context". Entomologia Experimentalis et Applicata. 97 (3): 237–249. doi:10.1046/j.1570-7458.2000.00736.x.

70. Allmann, S.; Baldwin, I. T. (2010). "Insects Betray Themselves in Nature to Predators by Rapid Isomerization of Green Leaf Volatiles". Science. 329 (5995): 1075–8. Bibcode:2010Sci...329.1075A. doi:10.1126/science.1191634. PMID 20798319.

71. Schuman, Meredith C.; Barthel, Kathleen; Baldwin, Ian T. (October 2012). "Herbivory-induced volatiles function as defenses increasing fitness of the native plant Nicotiana attenuata in nature" (PDF). eLife Sciences. 1. doi:10.7554/eLife.00007.

72. Heil, Martin; Brigitte Fiala, K. Eduard Linsenmair, Gerhard Zotz, Petra Menke (December 1997). "Food body production in Macaranga triloba (Euphorbiaceae): A plant investment in anti-herbivore defense via symbiotic ant partners". Journal of Ecology. British Ecological Society. 85 (6): 847–861. doi:10.2307/2960606. JSTOR 2960606.

73. Young, Truman P.; Cynthia H. Stubblefield; Lynne A. Isbell (January 1997). "Ants on swollen-thorn acacias: species coexistence in a simple system". Oecologia. 109 (1): 98–107. doi:10.1007/s004420050063.

74. Williams, Alan G.; Thomas G. Whitham (December 1986). "Premature Leaf Abscission: An Induced Plant Defense Against Gall Aphids". Ecology. Ecological Society of America. 67 (6): 1619–1627. doi:10.2307/1939093. JSTOR 1939093.

75. Lev-Yadun, Simcha; Amots Dafni; Moshe A. Flaishman; Moshe Inbar; Ido Izhaki; Gadi Katzir; Gidi Ne'eman (October 2004). "Plant coloration undermines herbivorous insect camouflage" (PDF). BioEssays. 26 (10): 1126–1130. doi:10.1002/bies.20112. PMID 15382135. Retrieved 2007-05-27.

76. Archetti, M., M (2000). "The origin of autumn colours by coevolution". J. Theor. Biol. 205 (4): 625–630. doi:10.1006/jtbi.2000.2089. PMID 10931756.

77. Hamilton, W. D.; Brown, S. P. (2001). "Autumn tree colours as a handicap signal" (PDF). Proc. R. Soc. B. 268 (1475): 1489–1493. doi:10.1098/rspb.2001.1672. PMC 1088768 . PMID 11454293.

78. Stamp, Nancy (March 2003). "Out of the quagmire of plant defense hypotheses". Quarterly Review of Biology. 78 (1): 23–55. doi:10.1086/367580. PMID 12661508.

79. Rhoades, D. F.; R. G. Cates. (1974). "Towards a general theory of plant antiherbivore chemistry". In V. C. Runeckles; E. E. Conn. Recent advances in phytochemistry: proceedings of the annual meeting of the Phytochemical society of North America. Boston: Academic Press. pp. 168–213. ISBN 0-12-612408-6.

80. Wilf, Peter; Conrad C. Labandeira; Kirk R. Johnson; Phyllis D. Coley; Asher D. Cutter (2001). "Insect herbivory, plant defense, and early Cenozoic climate

change" (PDF). Proceedings of the National Academy of Sciences. 98 (11): 6221–6226. Bibcode:2001PNAS...98.6221W. doi:10.1073/pnas.111069498. PMC 33449 . PMID 11353840. Retrieved 2007-05-27.

81. Feeny, P. (1976). "Plant apparency and chemical defense.". In James W. Wallace; Richard L. Mansell. Biochemical interaction between plants and insects: proceedings of the fifteenth annual meeting of the Phytochemical Society of North America. New York: Plenum Press. pp. 1–40. ISBN 0-306-34710-5.
82. D., McKey (1979). "The distribution of secondary compounds within plants.". In Gerald A. Rosenthal; Daniel H. Janzen. Herbivores, their interaction with secondary plant metabolites. Boston: Academic Press. pp. 55–133. ISBN 0-12-597180-X.
83. Krischik, V. A.; R. F. Denno (1983). "Individual, population, and geographic patterns in plant defense.". In Robert F. Denno; Mark S. McClure. Variable plants and herbivores in natural and managed systems. Boston: Academic Press. pp. 463–512. ISBN 0-12-209160-4.
84. Zangerl, Arthur R.; Claire E. Rutledge (April 1996). "The probability of attack and patterns of constitutive and induced defense: A test of optimal defense theory". The American Naturalist. 147 (4): 599–608. doi:10.1086/285868. JSTOR 2463237. Retrieved 2014-01-15.
85. Swain, Elisabeth; Chun Ping Li; Jonathan E. Poulton (1992). "Development of the Potential for Cyanogenesis in Maturing Black Cherry (Prunus serotina Ehrh.) Fruits". Plant Physiology. 98 (4): 1423–1428. doi:10.1104/pp.98.4.1423. PMC 1080367 . PMID 16668810.
86. Witmer, M.C. (1998). "Ecological and evolutionary implications of energy and protein requirements of avian frugivores eating sugary diets". Physiological Zoology. 71 (6): 599–610. PMID 9798248.
87. Pennings, Steven C.; Erin L. Siska; Mark D. Bertness (May 2001). "Latitudinal differences in plant palatability in Atlantic coast salt marshes". Ecology. Ecological Society of America. 82 (5): 1344–1359. doi:10.2307/2679994. JSTOR 2679994.
88. Bryant, John P.; Stuart Chapin, III; David R. Klein (May 1983). "Carbon/nutrient balance of boreal plants in relation to vertebrate herbivory". Oikos. Blackwell Publishing. 40 (3): 357–368. doi:10.2307/3544308. JSTOR 3544308.
89. Tuomi, J.; P. Niemela; F. S. Chapin, III; J. P. Bryant; S. Siren. (1988). "Defensive responses of trees in relation to their carbon/nutrient balance.". In William J. Mattson; Jean Levieux; C. Bernard-Dagan. Mechanisms of woody plant defenses against insects: search for pattern. Berlin: Springer-Verlag. pp. 57–72. ISBN 0-387-96673-0.
90. Colley, Phyllis D.; John P. Bryant; F. Stuart Chapin III (1985). "Resource availability and plant antiherbivore defense". Science. 230 (4728): 895–899. Bibcode:1985Sci...230..895C. doi:10.1126/science.230.4728.895. PMID 17739203.
91. Chapin, F. Stuart, III (1980). "The Mineral Nutrition of Wild Plants". Annual Review of Ecology and Systematics. 11: 233–260. doi:10.1146/annurev.

es.11.110180.001313. JSTOR 2096908. Retrieved 2014-01-15.

92. Fine, Paul V. A.; Italo Mesones; Phyllis D. Coley (July 2004). "Herbivores promote habitat specialization by trees in Amazonian forests". Science. 305 (5684): 663–5. Bibcode:2004Sci...305..663F. doi:10.1126/science.1098982. PMID 15286371.
93. Loomis, W. E. (1981). "Growth and differentiation—an introduction and summary.". In P. F. Wareing; I. D. J. Phillips. Growth and differentiation in plants. New York: Pergamon Press. pp. 1–17. ISBN 0-08-026351-8.
94. Herms, Daniel A.; William J. Mattson (September 1992). "The dilemma of plants: to grow or defend". Quarterly Review of Biology. 67 (3): 283–335. doi:10.1086/417659. JSTOR 2830650. Retrieved 2014-01-15.
95. Polavarapu, Sridhar (2001). "Plant Resistance to insects". Agricultural Entomology & Pest Management. Rutgers University. Retrieved 2007-05-16.
96. Painter, Reginald Henry (1951). Insect Resistance in Crop Plants. Lawrence: University of Kansas Press. OCLC 443998.
97. Chesnokov, Pavel G. (1953). Methods of Investigating Plant Resistance to Pests. Jerusalem: Israel Program for Scientific Translations. OCLC 3576157.
98. Gleadow, Roslyn M.; Ian E. Woodrow (2002). "Constraints on effectiveness of cyanogenic glycosides in herbivore defense". Journal of Chemical Ecology. 28 (7): 1301–13. doi:10.1023/A:1016298100201. PMID 12199497.
99. Vough, Lester R.; E. Kim Cassel (July 2002). "Prussic Acid Poisoning of Livestock: Causes and Prevention (ExEx 4016)" (PDF). Extension Extra. South Dakota State University Extension Service.
100. Grant, G; Linda J. More; Norma H. McKenzie; Arpad Pusztai (1982). "The effect of heating on the haemagglutinating activity and nutritional properties of bean (Phaseolus vulgaris) seeds". Journal of the Science of Food and Agriculture. 33 (12): 1324–6. doi:10.1002/jsfa.2740331220. PMID 7166934.
101. Tu Jean-Louis (1999). "Natural Toxins in Raw Foods and How Cooking Affects Them". Is Cooked Food Poison?. Beyond Vegetarianism. Retrieved 2007-05-22.
102. George W. (2004). The Pesticide Book. Willoughby: MeisterPro. ISBN 1-892829-11-8. firat Ware
103. Michael Smith, C. (2005). Plant Resistance to Arthropods: Molecular and Conventional Approaches. Berlin: Springer. ISBN 1-4020-3701-5.
104. Dethier, V. G. (March 1954). "Evolution of feeding preferences in phytophagous insects". Evolution. Society for the Study of Evolution. 8 (1): 33–54. doi:10.2307/2405664. JSTOR 2405664.
105. Russ, Karen. "Less toxic insecticides" (PDF). Clemson University Home & Garden Information Center. Retrieved 2007-05-27.
106. "Linden, J., Stoner, R., Knutson, K. Gardner-Hughes, C. "Organic Disease Control Elicitors". Agro Food Industry Hi-Te (p12-15 Oct 2000)" (PDF).

107. van Emden, H.F. (November 1999). "Transgenic Host Plant Resistance to Insects—Some Reservations". Annals of the Entomological Society of America. 92 (6): 788–797. Retrieved 2007-05-27.

108. Ghosh, B. (2000). "Polyamines and plant alkaloids". Indian Journal of Experimental Biology. 38 (11): 1086–91. PMID 11395950.

109. Eisner, Thomas (March 1990). "Prospecting for nature's chemical riches". Chemoecology. 1 (1): 38–40. doi:10.1007/BF01240585.

110. Albert T. Sneden. "The Quinine Alkaloids" (PDF). Medicinal Chemistry and Drug Design. Retrieved 2007-05-23.

111. Albert T. Sneden. "The Tropane Alkaloids" (PDF). Medicinal Chemistry and Drug Design. Retrieved 2007-05-23.

112. Albert T. Sneden. "Taxol (Paclitaxe)" (PDF). Medicinal Chemistry and Drug Design. Retrieved 2007-05-23.

113. Dangl, J. L.; Horvath, D. M.; Staskawicz, B. J. (2013). "Pivoting the Plant Immune System from Dissection to Deployment". Science. 341 (6147): 746. doi:10.1126/science.1236011.

114. Agrios, George N. (2005). Plant Pathology, Fifth Edition. Academic Press. ISBN 978-0120445653.

115. Lutz, Diana (2012). Key part of plants' rapid response system revealed. Washington University in St. Louis.

116. Dadakova, K.; Havelkova, M.; Kurkova, B.; Tlolkova, I.; Kasparovsky, T.; Zdrahal, Z.; Lochman, J. (2015-04-24). "Proteome and transcript analysis of Vitis vinifera cell cultures subjected to Botrytis cinerea infection". Journal of Proteomics. 119: 143–153. doi:10.1016/j.jprot.2015.02.001.

117. Dadakova, K.; Havelkova, M.; Kurkova, B.; Tlolkova, I.; Kasparovsky, T.; Zdrahal, Z.; Lochman, J. (2015-04-24). "Proteome and transcript analysis of Vitis vinifera cell cultures subjected to Botrytis cinerea infection". Journal of Proteomics. 119: 143–153. doi:10.1016/j.jprot.2015.02.001.

118. Dodds, P. N.; Rathjen, J. P. (2010). "Plant immunity: Towards an integrated view of plant–pathogen interactions". Nature Reviews Genetics. 11 (8): 539. doi:10.1038/nrg2812.

119. Jones, J. D.; Dangl, J.L. (2006). "The plant immune system". Nature. 444 (7117): 323–329. Bibcode:2006Natur.444..323J. doi:10.1038/nature05286. PMID 17108957.

120. Li, B.; Meng, X.; Shan, L.; He, P. (2016). "Transcriptional Regulation of Pattern-Triggered Immunity in Plants". Cell Host Microbe. 19 (5): 641–650. doi:10.1016/j.chom.2016.04.011. PMID 27173932.

121. Nurnberger, T.; Brunner, F., Kemmerling, B., and Piater, L.; Kemmerling, B; Piater, L (2004). "Innate immunity in plants and animals: striking similarities and obvious differences". Immunological Reviews. 198: 249–266. doi:10.1111/j.0105-2896.2004.0119.x. PMID 15199967.

122. Thomma, B.; Nurnberger, T.; Joosten, M. (2011). "Of PAMPs and Effectors:

The Blurred PTI-ETI Dichotomy". The Plant Cell. 23 (4): 15. doi:10.1105/tpc.110.082602. PMC 3051239 . PMID 21278123.

123. Lindeberg, M; Cunnac S, Collmer A (2012). "Pseudomonas syringae type III effector repertoires: last words in endless arguments". Trends Microbiol. 20: 199–208. doi:10.1016/j.tim.2012.01.003. PMID 22341410.

124. Rafiqi, M; Ellis JG, Ludowici VA, Hardham AR, Dodds PN (2012). "Challenges and progress towards understanding the role of effectors in plant-fungal interactions". Curr Opin Plant Biol. 15: 477–482. doi:10.1016/j.pbi.2012.05.003. PMID 22658704.

125. Hewezi, T; Baum T (2013). "Manipulation of plant cells by cyst and root-knot nematode effectors". Mol Plant Microbe Interact. 26: 9–16. doi:10.1094/MPMI-05-12-0106-FI. PMID 22809272.

126. Hammond-Kosack KE; Parker JE (Apr 2003). "Deciphering plant-pathogen communication: fresh perspectives for molecular resistance breeding.". Curr Opin Biotechnol. 14 (2): 177–193. doi:10.1016/s0958-1669(03)00035-1. PMID 12732319.

127. Dadakova, Katerina; Klempova, Jitka; Jendrisakova, Tereza; Lochman, Jan; Kasparovsky, Tomas (2013-12-01). "Elucidation of signaling molecules involved in ergosterol perception in tobacco". Plant Physiology and Biochemistry. 73: 121–127. doi:10.1016/j.plaphy.2013.09.009.

128. Moore, J. W.; Loake, G. J.; Spoel, S. H. (12 August 2011). "Transcription Dynamics in Plant Immunity". The Plant Cell. 23 (8): 2809–2820. doi:10.1105/tpc.111.087346. PMC 3180793 . PMID 21841124.

129. Sadanandom, Ari; Bailey, Mark; Ewan, Richard; Lee, Jack; Nelis, Stuart (1 October 2012). "The ubiquitin-proteasome system: central modifier of plant signalling". New Phytologist. 196 (1): 13–28. doi:10.1111/j.1469-8137.2012.04266.x. PMID 22897362.

130. Trujillo, M; Shirasu, K (August 2010). "Ubiquitination in plant immunity.". Current Opinion in Plant Biology. 13 (4): 402–8. doi:10.1016/j.pbi.2010.04.002. PMID 20471305.

131. Craig, A.; Ewan, R.; Mesmar, J.; Gudipati, V.; Sadanandom, A. (10 March 2009). "E3 ubiquitin ligases and plant innate immunity". Journal of Experimental Botany. 60 (4): 1123–1132. doi:10.1093/jxb/erp059. PMID 19276192.

132. Moon, J. (1 December 2004). "The Ubiquitin-Proteasome Pathway and Plant Development". The Plant Cell Online. 16 (12): 3181–3195. doi:10.1105/tpc.104.161220.

133. Trujillo, Marco; Shirasu, Ken (1 August 2010). "Ubiquitination in plant immunity". Current Opinion in Plant Biology. 13 (4): 402–408. doi:10.1016/j.pbi.2010.04.002. PMID 20471305.

134. Shirsekar, Gautam; Dai, Liangying; Hu, Yajun; Wang, Xuejun; Zeng, Lirong; Wang, Guo-Liang; Hu, Yajun; Wang, Xuejun; Zeng, Lirong; Wang, Guo-Liang (2010). "Role of Ubiquitination in Plant Innate Immunity and Pathogen Virulence". Journal of Plant Biology. 53 (1): 10–18. doi:10.1007/s12374-009-9087-x.

135. Stuthman, DD; Leonard, KJ; Miller-Garvin, J (2007). "Breeding crops for durable resistance to disease". Advances in Agronomy: 319–367.

136. Tabashnik, Bruce E.; Brevault, Thierry; Carriere, Yves (2013). "Insect resistance to Bt crops: lessons from the first billion acres". Nature Biotechnology. 31: 510–521. doi:10.1038/nbt.2597.

137. Kavanagh, T. A.; Spillane, C. (1995-02-01). "Strategies for engineering virus resistance in transgenic plants". Euphytica. 85 (1-3): 149–158. doi:10.1007/BF00023943. ISSN 0014-2336.

138. Kaniewski, Wojciech K.; Thomas, Peter E. (2004). "The Potato Story". AgBioForum. 7 (1&2): 41–46.

139. Lacombe et al., 2010 Interfamily transfer of a plant pattern-recognition receptor confers broad-spectrum bacterial resistance, Nature Biotech 28, 365–369

140. Karthikeyan, A.; Deivamani, M.; Shobhana, V. G.; Sudha, M.; Anandhan, T. (2013). "RNA interference: Evolutions and applications in plant disease management". Archives of Phytopathology and Plant Protection. 46 (12): 1430. doi:10.1080/03235408.2013.769315.

141. Bettgenhaeuser, J; Gilbert B, Ayliffe M, Moscou MJ (2014). "Nonhost resistance to rust pathogens - a continuation of continua". Front Plant Sci. 5: 664. doi:10.3389/fpls.2014.00664. PMID 25566270.

142. Restrepo, S; Tabima JF, Mideros MF, Grünwald NJ, Matute DR (2014). "Speciation in fungal and oomycete plant pathogens". Annu Rev Phytopathol. 52: 289–316. doi:10.1146/annurev-phyto-102313-050056. PMID 24906125.

143. McDonald, B. A.; Linde, C (2002). "Pathogen population genetics, evolutionary potential, and durable resistance". Annual Review of Phytopathology. 40: 349–79. doi:10.1146/annurev.phyto.40.120501.101443. PMID 12147764.

CHAPTER 2

Plant Disease Epidemiology

Chapter Contents

Epidemiology is the science of disease in populations. It is a quantitative discipline with strong conceptual foundations and practical applications. Epidemics of plant diseases have had major impacts on agricultural and horticultural crops and on forest-tree populations with major socio-economic and political consequences.

INTRODUCTION

Plant disease epidemiology is the study of disease in plant populations. Much like diseases of humans and other animals, plant diseases occur due to pathogens such as bacteria, viruses, fungi, oomycetes, nematodes, phytoplasmas, protozoa, and parasitic plants. Plant disease epidemiologists strive for an understanding of the cause and effects of disease and develop strategies to intervene in situations where crop losses may occur. Typically successful intervention will lead to a low enough level of disease to be acceptable, depending upon the value of the crop.

Plant disease epidemiology is often looked at from a multi-disciplinary approach, requiring biological, statistical, agronomic and ecological perspectives. Biology is necessary for understanding the pathogen and its life cycle. It is also necessary for understanding the physiology of the crop and how the pathogen is adversely affecting it. Agronomic practices often influence disease incidence for better or for worse. Ecological influences are numerous. Native species of plants may serve as reservoirs for pathogens that cause disease in crops. Statistical models are often applied in order to summarize and describe the complexity of plant disease epidemiology, so that disease processes can be more readily understood. For example, comparisons between patterns of disease progress for different diseases, cultivars, management strategies, or environmental settings can help in determining how plant diseases may best be managed. Policy can be influential in the occurrence of diseases, through actions such as restrictions on imports from sources where a disease occurs.

In 1963 J. E. van der Plank published "Plant Diseases: Epidemics and Control", a seminal work that created a theoretical framework for the study of the epidemiology of plant diseases. This book provides a theoretical framework based on experiments in many different host pathogen systems and moved the study of plant disease epidemiology forward rapidly, especially for fungal foliar pathogens. Using this framework we can now model and determine thresholds for epidemics that take place in a homogeneous environment such as a mono-cultural crop field.

Elements of an Epidemic

Disease epidemics in plants can cause huge losses in yield of crops as well threatening to wipe out an entire species such as was the case with Dutch elm disease and could occur with Sudden Oak Death. An epidemic of potato late blight, caused by *Phytophthora infestans*, led to the Great Irish Famine and the loss of many lives.

Commonly the elements of an epidemic are referred to as the "disease triangle": a susceptible host, pathogen, and conducive environment. For disease to occur all three of these must be present. Below is an illustration of this point. Where all three items meet there is disease. The fourth element missing from this illustration for an epidemic to occur, is time. As long as all three of these elements are present disease can initiate, an epidemic will only ensue if all three continue to be present. Any one of the three might be removed from the equation though. The host might out-grow susceptibility as with high-temperature adult-plant resistance, the environment changes and is not conducive for the pathogen to cause disease, or the pathogen is controlled through a fungicide application for instance.

Sometimes a fourth factor of time is added as the time at which a particular infection occurs, and the length of time conditions remain viable for that infection, can also play an important role in epidemics. The age of the plant species can also play a role, as certain species change in their levels of disease resistance as they mature; a process known as ontogenic resistance.

If all of the criteria are not met, such as a susceptible host and pathogen are present but the environment is not conducive to the pathogen infecting and causing disease, disease cannot occur. For example, corn is planted into a field with corn residue that has the fungus *Cercospora zea-maydis*, the causal agent of Grey leaf spot of corn, but if the weather is too dry and there is no leaf wetness the spores of the fungus in the residue cannot germinate and initiate infection.

Likewise, it stands to reason if the host is susceptible and the environment favours the development of disease but the pathogen is not present there is no disease. Taking the example above, the corn is planted into a ploughed field where there is no corn residue with the fungus *Cercospora zea-maydis*, the causal agent of Grey leaf spot of corn, present but the weather means long periods of leaf wetness, there is no infection initiated.

When a pathogen requires a vector to be spread then for an epidemic to occur the vector must be plentiful and active.

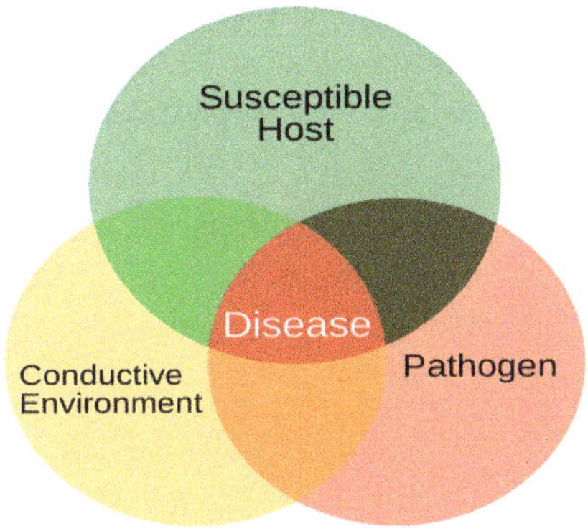

Plant disease triangle illustration

Types of Epidemics

Monocyclic epidemics are caused by pathogens with a low birth rate and death rate, meaning they only have one infection cycle per season. They are typical of soil-borne diseases such as Fusarium wilt of flax. Polycyclic epidemics are caused by pathogens capable of several infection cycles a season. They are most often caused by airborne diseases such as powdery mildew. Bimodal polycyclic epidemics can also occur. For example, in brown rot of stone fruits the blossoms and the fruits are infected at different times.

For some diseases it is important to consider the disease occurrence over several growing seasons, especially if growing the crops in monoculture year after year or growing perennial plants. Such conditions can mean that the inoculum produced in one season can be carried over to the next leading to a build of an inoculum over the years. In the tropics there are no clear-cut breaks between growing seasons as there are in temperate regions and this can lead to accumulation of inoculum.

Epidemics that occur under these conditions are referred to as *polyetic* epidemics and can be caused by both monocyclic and polycyclic pathogens. Apple powdery mildew is an example of a polyetic epidemic caused by a polycyclic pathogen and Dutch Elm disease a polyetic epidemic caused by a monocyclic pathogen.

Employment

Plant disease epidemiologists are typically employed as researchers by universities, or governmental institutions such as the USDA. However, private companies in agricultural fields also employ epidemiologists.

PLANT DISEASE FORCASTING

Plant disease forecasting is a management system used to predict the occurrence or change in severity of plant diseases. At the field scale, these systems are used by growers to make economic decisions about disease treatments for control. Often the systems ask the grower a series of questions about the susceptibility of the host crop, and incorporate current and forecast weather conditions to make a recommendation. Typically a recommendation is made about whether disease treatment is necessary or not. Usually treatment is a pesticide application.

Forecasting systems are based on assumptions about the pathogen›s interactions with the host and environment, the disease triangle. The objective is to accurately predict when the three factors – host, environment, and pathogen – all interact in such a fashion that disease can occur and cause economic losses.

In most cases the host can be suitably defined as resistant or susceptible, and the presence of the pathogen may often be reasonably ascertained based on previous cropping history or perhaps survey data. The environment is usually the factor that controls whether disease develops or not. Environmental conditions may determine the presence of the pathogen in a particular season through their effects on processes such as overwintering. Environmental conditions also affect the ability of the pathogen to cause disease, e.g. a minimum leaf wetness duration is required for grey leaf spot of corn to occur. In these cases a disease forecasting system attempts to define when the environment will be conducive to disease development.

Good disease forecasting systems must be reliable, simple, cost-effective and applicable to many diseases. As such they are normally only designed for diseases that are irregular enough to warrant a prediction system, rather than diseases that occur every year for which regular treatment should be employed. Forecasting systems can only be designed if there is also an understanding on the actual disease triangle parameters.

Examples of Disease Forecasting Systems

Forecasting systems may use one of several parameters in order to work out disease risk, or a combination of factors. One of the first forecasting systems designed was for Stewart's wilt and based on winter temperature index as low temperatures would kill the vector of the disease so there would be no outbreak. An example of a multiple disease/pest forecasting system is the EPIdemiology, PREdiction, and PREvention (EPIPRE) system developed in the Netherlands for winter wheat that focused on multiple pathogens. USPEST.org graphs risks of various plants diseases based on weather forecasts with hourly resolution of leaf wetness. Forecasting models are often based on a relationship like simple linear regression where x is used to predict y. Other relationships can be modelled using population growth curves. The growth curve that is used will depend on the nature of the epidemic. Polycyclic epidemics such as potato late blight are usually best modelled by using the logistic model, whereas monocyclic epidemics may be best modelled using the monomolecular model. Correct choice of a model is essential for a disease forecasting system to be useful.

Plant disease forecasting models must be thoroughly tested and validated after being developed. Interest has arisen lately in model validation through the quantification of the economic costs of false positives and false negatives, where disease prevention measures may be used when unnecessary or not applied when needed respectively. The costs of these two types of errors need to be weighed carefully before deciding to use a disease forecasting system.

Future Developments

In the future, disease forecasting systems may become more useful as computing power increases and the amount of data that is available to plant pathologists to construct models increases. Good forecasting systems also may become increasingly important with climate change. It will be important to be able to accurately predict where disease outbreaks may occur, since they may not be in the historically known areas.

REFERENCES

1. Agrios, George (2005). Plant Pathology. Academic Press. ISBN 978-0-12-044565-3.
2. Campbell, C. L.; Madden, L. V. (1990). Introduction to Plant Disease Epidemiology. New York: Wiley and Sons. ISBN 0-471-83236-7.
3. Esker, P. D.; A.H. Sparks; L. Campbell; Z. Guo; M. Rouse; S.D. Silwal; S. Tolos; B. Van Allen; K.A. Garrett. "Ecology and Epidemiology in R: Disease Forecasting". The Plant Health Instructor. APS Press. doi:10.1094/PHI-A-2008-01.
4. APS Education Centre - Stewart's wilt of corn
5. Reinink, K (1986). "Experimental verification and development of EPIPRE, a supervised disease and pest management system for wheat". European Journal of Plant Pathology. SpringerLink. 92 (1): 3–14. doi:10.1007/BF01976371.
6. Madden, Laurence; Gareth Hughes; Frank Van Den Bosch (2007). Study of Plant Disease Epidemics. American Phytopathological Society. ISBN 978-0-89054-354-2.
7. Agrios, George (2005). Plant Pathology. Academic Press. ISBN 978-0-12-044565-3.
8. Arneson, PA (2001). "Plant disease epidemiology: temporal aspects". Plant Health Instructor. American Phytopathological Society. doi:10.1094/PHI-A-2001-0524-01.
9. Madden, Laurence; Gareth Hughes; Frank Van Den Bosch (2007). Study of Plant Disease Epidemics. American Phytopathological Society. ISBN 978-0-89054-354-2.

10. Drenth, A (2004). "Fungal epidemics – does spatial structure matter?". New Phytologist. Blackwells. 163 (1): 4–7. doi:10.1111/j.1469-8137.2004.01116.x.

11. Cormac Ó Gráda, Ireland's Great Famine, University College Dublin, 2006, ISBN 978-1-9045-5858-3, p. 7

12. Shultz, T.R; Line, R.F (1992). "High-Temperature, Adult-Plant Resistance to Wheat Stripe Rust and Effects on Yield Components". Agronomy Journal. American Society of Agronomy. 84 (2): 170–175. doi:10.2134/agronj1992.00021962008400020009x.

CHAPTER 3

Plant Viruses

Chapter Contents

- **Introduction**
- **Tospovirus**
- **Pepper Mild Mottle Virus**

Viruses are infectious pathogens that are too small to be seen with a light microscope, but despite their small size they can cause chaos. The simplest viruses are composed of a small piece of nucleic acid surrounded by a protein coat. As is the case with other organisms, viruses carry genetic information in their nucleic acid which typically specifies three or more proteins. All viruses are obligate parasites that depend on the cellular machinery of their hosts to reproduce. Viruses are not active outside of their hosts, and this has led some people to suggest that they are not alive. All types of living organisms including animals, plants, fungi, and bacteria are hosts for viruses, but most viruses infect only one type of host. Viruses cause many important plant diseases and are responsible for losses in crop yield and quality in all parts of the world.

INTRODUCTION

Plant viruses are viruses that affect plants. Like all other viruses, plant viruses are obligate intracellular parasites that do not have the molecular machinery to replicate without a host. Plant viruses are pathogenic to higher plants.While this does not intend to list all plant viruses, it discusses some important viruses as well as their uses in plant molecular biology

Pepper mild mottle virus

Leaf curl virus

Although plant viruses are not nearly as well understood as the animal counterparts, one plant virus has become iconic. The first virus to be discovered (see below) was *Tobacco mosaic virus* (TMV). This and other viruses cause an estimated US$60 billion loss in crop yields worldwide each year. Plant viruses are grouped into 73 genera and 49 families. However, these figures relate only to cultivated plants that represent only a tiny fraction of the total number of plant species. Viruses in wild plants have been poorly studied, but those studies that exist almost overwhelmingly show that such interactions between wild plants and their viruses do not appear to cause disease in the host plants.

To transmit from one plant to another and from one plant cell to another, plant viruses must use strategies that are usually different from animal viruses. Plants do not move, and so plant-to-plant transmission usually involves vectors (such as insects). Plant cells are surrounded by solid cell walls, therefore transport through plasmodesmata is the preferred path for virions to move between plant cells. Plants probably have specialized mechanisms for transporting mRNAs through plasmodesmata, and these mechanisms are thought to be used by RNA viruses to spread from one cell to another.

Plant defenses against viral infection include, among other measures, the use of siRNA in response to dsRNA. Most plant viruses encode a protein to suppress this response. Plants also reduce transport through plasmodesmata in response to injury.

The discovery of plant viruses causing disease is often accredited to A. Mayer (1886) working in the Netherlands demonstrated that the sap of mosaic obtained from tobacco leaves developed mosaic symptom when injected in healthy plants. However the infection of the sap was destroyed when it was boiled. He thought that the causal agent was the bacteria. However, after larger inoculation with a large number of bacteria, he failed to develop a mosaic symptom.

In 1898, Martinus Beijerinck, who was a Professor of Microbiology at the Technical University the Netherlands, put forth his concepts that viruses were small and determined that the «mosaic disease» remained infectious when passed through a Chamberland filter-candle. This was in contrast to bacteria microorganisms, which were retained by the filter. Beijerinck referred to the infectious filtrate as a "contagium vivum fluidum», thus the coinage of the modern term «virus».

After the initial discovery of the 'viral concept' there was need to classify any other known viral diseases based on the mode of transmission even though microscopic observation proved fruitless. In 1939 Holmes published a classification list of 129 plant viruses. This was expanded and in 1999 there were 977 officially recognized, and some provisional, plant virus species.

The purification (crystallization) of TMV was first performed by Wendell Stanley, who published his findings in 1935, although he did not determine that the RNA was the infectious material. However, he received the Nobel Prize in Chemistry in 1946. In the 1950s a discovery by two labs simultaneously proved that the purified RNA of the TMV was infectious which reinforced the argument. The RNA carries genetic information to code for the production of new infectious particles.

More recently virus research has been focused on understanding the genetics and molecular biology of plant virus genomes, with a particular interest in determining how the virus can replicate, move and infect plants. Understanding the virus genetics and protein functions has been used to explore the potential for commercial use by biotechnology companies. In particular, viral-derived sequences have been used to provide an understanding of novel forms of resistance. The recent boom in technology allowing humans to manipulate plant viruses may provide new strategies for production of value-added proteins in plants.

Structure

Viruses are extremely small and can only be observed under an electron microscope. The structure of a virus is given by its coat of proteins, which surround the viral genome. Assembly of viral particles takes place spontaneously.

Over 50% of known plant viruses are rod-shaped (flexuous or rigid). The length of the particle is normally dependent on the genome but it is usually between 300–500 nm with a diameter of 15–20 nm. Protein subunits can be placed around the circumference of a circle to form a disc. In the presence of the viral genome, the discs are stacked, then a tube is created with room for the nucleic acid genome in the middle.

The second most common structure amongst plant viruses are isometric particles. They are 25–50 nm in diameter. In cases when there is only a single coat protein, the basic structure consists of 60 T subunits, where T is an integer. Some viruses may have 2 coat proteins that associate to form an icosahedral shaped particle.

There are three genera of *Geminiviridae* that possess geminate particles which are like two isometric particles stuck together.

A very small number of plant viruses have, in addition to their coat proteins, a lipid envelope. This is derived from the plant cell membrane as the virus particle buds off from the cell.

Transmission of Plant Viruses

Through Sap

Viruses can be spread by direct transfer of sap by contact of a wounded plant with a healthy one. Such contact may occur during agricultural practices, as by damage caused by tools or hands, or naturally, as by an animal feeding on the plant. Generally TMV, potato viruses and cucumber mosaic viruses are transmitted via sap.

Insects

Plant viruses need to be transmitted by a vector, most often insects such as leafhoppers. One class of viruses, the Rhabdoviridae, has been proposed to actually be insect viruses that have evolved to replicate in plants. The chosen insect vector of a plant virus will often be the determining factor in that virus's host range: it can only infect plants that the insect vector feeds upon. This was shown in part when the old world white fly made it to the United States, where it transferred many plant viruses into new hosts. Depending on the way they are transmitted, plant viruses are classified as non-persistent, semi-persistent and persistent. In non-persistent transmission, viruses become attached to the distal tip of the stylet of the insect and on the next plant it feeds on, it inoculates it with the virus. Semi-persistent viral transmission involves the virus entering the foregut of the insect. Those viruses that manage to pass through the gut into the haemolymph and then to the salivary glands are known as persistent. There are two sub-classes of persistent viruses: propagative and circulative. Propagative viruses are able to replicate in both the plant and the insect (and may have originally been insect viruses), whereas circulative can not. Circulative viruses are protected inside aphids by the chaperone protein symbionin, produced by bacterial symbionts. Many plant viruses encode within their genome polypeptides with domains essential for transmission by insects. In non-persistent and semi-persistent viruses, these domains are in the coat protein and another protein known as the helper component. A bridging hypothesis has been proposed to explain how these proteins aid in insect-mediated viral transmission. The helper component will bind to the specific domain of the coat protein, and then the insect mouthparts — creating a bridge. In persistent propagative viruses, such as tomato spotted wilt virus (TSWV), there is often a lipid coat surrounding the proteins that is not seen in other classes of plant viruses. In the case of TSWV, 2 viral proteins are expressed in this lipid envelope. It has been proposed that the viruses bind via these proteins and are then taken into the insect cell by receptor-mediated endocytosis.

TOSPOVIRUS

The tospoviruses are a genus (Tospovirus) of negative RNA virus found within the family Bunyaviridae. They are the sole group of plant infecting viruses in this family, as all other described members of the Bunyaviridae infect animals. The genus takes its name from the discovery of tomato spotted wilt virus (TSWV) in Australia in 1915. It remained the only member of the family until the early 1990s when genetic characterisation of viruses discovered in plants became more common. There are now at least twenty viral species in the family with more being recorded and described on a relatively regular basis. Together, these viruses have been documented infecting over eight hundred different plant species from 82 different families.

A tomato infected with tomato spotted wilt virus.

Genome

This virus has a single stranded RNA genome with negative polarity, therefore it is classified as a Class V virus ((-)ssRNA viruses). The structure of the genome resembles that of the genus Phlebovirus. The genome is linear and is 17.2 kb in size. It is segmented into three segments termed S (2.9kb), M (5.4kb) and L (8.9kb). The M and S RNA segments encode for proteins in an ambisense orientation.

Transmission

Tospoviruses are arboviruses usually vectored by a thrips. At least ten species of thrips belonging to family Thripidae have been confirmed as vectors for the transmission of thirteen or more tospoviruses. The thrips vectors are not closely related, implying an independent origin of infection for each thrips, possibly transmitted horizontally through shared hosts. There may be other species of thrips competent to transmit similar viruses, but they have not been documented on crops of economic significance.

Recent research concludes that thrips can only be infected by tospovirus during the larval phases of development, as pupation and metamorphosis separate the connection between the salivary glands and the infected muscle tissue of the mid-gut. Adults transmit the virus from infected salivary glands, and uninfected adults will not transmit the virus. Obviously, controlling the infection by limiting transmission from infected plants to larval thrips or by preventing adult dispersal from infected plants are key strategies in preventing an epidemic of the disease.

Pepper infected with Tomato spotted wilt virus. Plant of Pimiento L sweet pepper.

Agricultural Importance

Infection with the virus results in spotting and wilting of the plant, reduced vegetative output, and eventually death. No antiviral cures have been developed for plants infected with a Tospovirus, and infected plants should be removed from a field and destroyed in order to prevent the spread of the disease.

There are a large number of plant families that are known to be affected by viruses of the Tospovirus genus. These include both food crops (such as peanuts, watermelons, capsicums, tomatoes, zucchinis, *et al.*) as well as ornamental species which are important to flower farms (calla lily, impatiens, chrysanthemums, iris, *et al.*). For a more complete list of hosts examine the Tospovirus host list at Kansas State University.

Diagnosis

Early symptoms of infection are difficult to diagnose. In young infected plants the characteristic symptoms consist of inward cupping of leaves and leaves that develop a bronze cast followed by dark spots. As the infection progresses additional symptoms develop which include dark streaks on the main stem and wilting of the top portion of the plant. Fruit may be deformed, show uneven ripening and often have raised bumps on the surface. Once a plant becomes infected the disease cannot be controlled.

Serological and molecular tests are commercially available to diagnose TSWV as well as a second common tospovirus found in ornamentals, *Impatiens necrotic spot virus* (INSV). Cytological studies of TSWV and INSV have shown that these viruses produce granular inclusions in the cytoplasm of infected plants. These inclusions can be seen in the light microscope with proper staining techniques. These inclusions can be diagnostic.

Epidemiology

TSWV is prevalent in warm climates in regions with a high population of thrips. This virus is an agricultural pest in Asia, America, Europe and Africa. Over the past 15 years outbreaks of Tomato spotted wilt disease have become more prevalent in these regions. Therefore TSWV is described as an emerging viral disease of plants. The increased prevalence is largely because of the successful survival of the thrips vector *Frankliniella occidentalis*. Another thrips, *Scirtothrips dorsalis*, has also been implicated in the transmission of at least three tospoviruses, but there remains some controversy over its efficiency as a vector. Immunological testing and vector-competence studies suggest that *S. dorsalis* may represents a non-transmitting carrier for some strains of virus.

The success of this virus has also been attributed to the acquisition of a gene in the M segment of the genome which encodes a movement protein. This protein allows the virus to infect a wide range of hosts. The gene encoding this protein was likely acquired by recombination from either a plant host or from another plant virus.

Management

Control of this disease is difficult. One of the reasons for this is that the wide host range allows the virus to successfully overseason from one crop to the next. To prevent spread of the virus infected plants should be immediately removed away from neighbouring plants. Control of insects, especially thrips, is important to reduce spread of the virus by vectors.

PEPPER MILD MOTTLE VIRUS

Pepper mild mottle virus (PMMoV) is a plant pathogenic virus that occurs worldwide on species of field grown bell, hot and ornamental pepper species. It is caused by members of the plant virus genus Tobamovirus- otherwise known as the Tobacco mosaic virus family. Tobamovirus are viruses that contain positive sense RNA genomes that infect plants. Symptoms of the disease vary depending on the cultivar. Typical symptoms include the chlorosis of leaves, stunting, and distorted and lumpy fruiting structures. The virus is spread by mechanical transmission and infected seeds. Avoidance is the best means of controlling the disease because once a plant is infected it cannot be treated. Only seeds that have been tested and treated for the pathogen should be planted.

The origin of PMMoV has been linked to the Tomato mosaic virus, as they both reside in the Tobacco mosaic virus family. The Tunisian Journal of Plant Protection brought about the link between PMMoV to ToMV from a French study dating back to 1964. ToMV affects a wide range of *Solanaceous* crops and a strain of this virus likely mutated into PMMoV.

Host and Symptoms

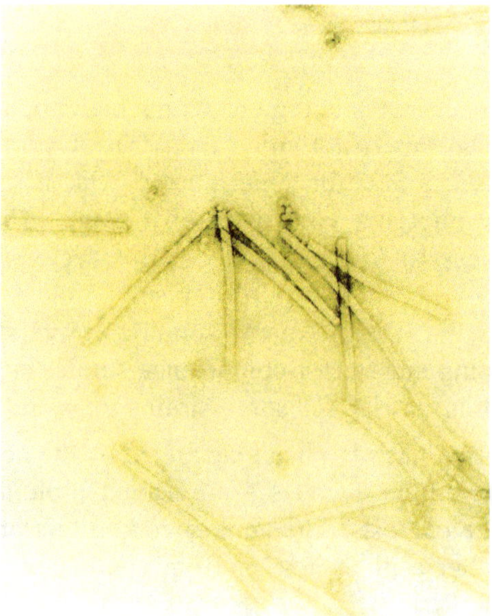

rigid rod shape of Tobamovirus

Healthy Peppers

The Pepper Mild Mottle Virus is the major viral pathogen of peppers (Capsicum spp.). The host range of PMMoV include most cultivars and species of pepper (Capsicum spp.). This virus strain does not infect tomato, eggplant, or tobacco; however, other members of the genus *Tobamovirus* can infect these other hosts PMMoV is one of at least 4 different species of *Tobamovirus* that infect peppers.The others include *Tobacco mosaic virus* (TMV), *Tomato mosaic virus* (ToMV) and *Tobacco mild green mosaic virus* (TMGMV).

Diseased Peppers

This pathogen is known to occur throughout the world and it frequently results in significant crop losses or reductions in both field and greenhouse plantings. The virus has been identified in places like Australia, Japan, China, Taiwan, Europe, and North Africa. Since 1997, PMMoV has been the cause of many major outbreaks in the southeastern states namely Georgia and Florida.

Symptoms caused by this pathogen vary based on the specific host cultivar; however, a majority of the symptoms are very similar between the different hosts. Symptoms usually include various degrees of mottling, chlorosis, curling, dwarfing, and distortion of the fruit, leaves, and even whole plants. The symptoms on fruit include: a reduction in size, mottling and color changes, and an obvious distorted and lumpy appearance. Also, many times brown necrotic streaks or splotches can be seen on the leaves and fruit. The symptoms can easily be seen on new growth, and they are far more pronounced if the plant was infected when it was young rather than when it was older.

This disease is harmful because of the mild foliar symptoms (chlorosis, necrosis, etc.) and due to this many times the pathogen goes unnoticed until the more evident symptoms on the fruit appear. This is why there are higher yield losses because symptoms only become evident during the fruiting stage right before the crops are supposed to harvested.

Disease Transmission

PMMoV is a rigid rod shaped virus pathogen that can be easily transmitted to healthy plants via mechanical inoculation or contact between plants or another medium that can carry the pathogen such as hands, gloves, and clothing. The virus is commonly found on the outer seed coat and very rarely in the endosperm of seeds of the infected plants. Being seedborne, the virus primarily infects through mechanical contamination during seed transplanting and other agricultural procedures. It enters the plant/seed through microscopic abrasions or wounds. The virus is in vitro, which in this case means it can survive in an isolated environment, and being highly infectious it can easily be transmitted during normal crop maintenance. PMMoV is exceptionally stable and it is known to survive for extended periods of time in plant debris and on greenhouse structures, pots, and horticultural tools. There are no known insect vectors; however, humans can be considered the main vector for this pathogen.

Environment

This virus is widespread and is found almost anywhere peppers are grown, this is because the virus is most often introduced with the pepper seed. The virus moves long distances on the seed and moves short distances via plant-to-plant contact, handling of plants by contaminated implements and workers aids in the spread of the virus. Soils with low organic matter content, specifically humus will much more readily promote the adsorption of PMMoV. Specifically, the allophanic clays and iron minerals in these soils will promote the adsorption of PMMoV.

Disease incidence has occurred mainly in greenhouse environments and hot, humid environments like southwest and southeast Florida and in regions like North Africa, and Japan. This is an indication of the type of soil that is optimal for the virus; warm or hot and humid. Greenhouses, specifically, are an ideal production environment for rapid spread of the disease.

Prevention, Control and Management

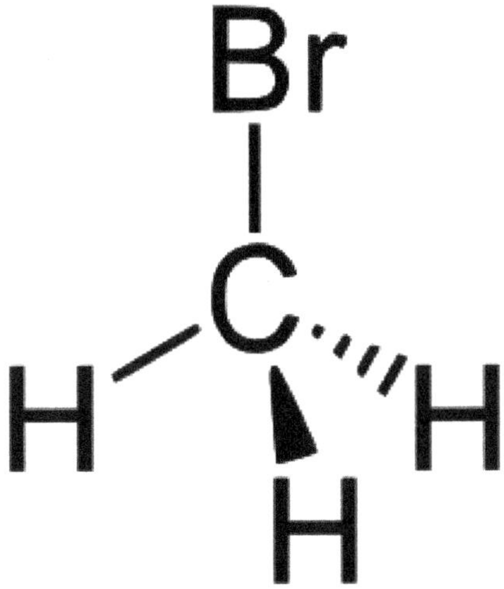

Structure of Methyl Bromide- Used until 2005 when it was banned by the Montreal Protocol

Up until 2005, the primary means of prevention was a pre-plant treatment of *Bromomethane* or methyl bromide. The import and production of this fumigation chemical was banned under the Montreal Protocol. A modern organic method of prevention was found by a Japanese study on soil adsorptions role in the adsorption of PMMoV. The study found that increased humus content in a soil will have an inhibitory effect on PMMoV adsorption.

As with all plant viruses, but especially with those in the *Tobamovirus* genus, avoidance is the best means of control. Growers must practice good sanitation procedures and only plant clean seeds. Growers must also use caution when handling plants, plants with abrasions or wounds give the virus an opportunity to enter the tissue of the plant.There are no chemical or biological control methods that can be used to control the disease once the plant is infected. Though often a challenge, accurate identification of PMMoV is key to successful control of the disease. (See reference 7 for a detailed description of control measures for this virus.)

Diseased plant material will remain infectious until completely broken down. Tillage, increased irrigation, and high temperatures encourage the breakdown of plant material in the soil. Any infected plant material in the soil can serve as a source of inoculum for subsequent crops so crop rotation should be practiced, if possible. Volunteer peppers and weeds, particularly those in the Solanaceae family (such as nightshades), should be removed to reduce possible sources of infections. Due to the fact that smoke can spread the disease, burial or composting of the diseased plants should be implemented rather than burning. The composted diseased plant remains should not be used on a pepper crop or other *Solanaceous* crops.

Importance

Controlling this virus is important for pepper production worldwide, but recent research shows that this plant disease may be transmitted to humans. A French study identified a local source of PMMoV and linked the presence of PMMoV RNA in stool with a specific immune response and clinical symptoms. Although clinical symptoms may be imputable to another co-factor, including spicy food, the data in the study suggested the possibility of a direct or indirect pathogenic role of plant viruses in humans.

In an article from the Applied and Environmental Microbiology Journal, it was found that PMMoV could be a potential indicator of fecal pollution. Their results demonstrate that PMMoV is widespread and abundant in wastewater from the United States, suggesting the utility of this virus as an indicator of human fecal pollution. Quantitative PCR was used to determine the abundance of PMMoV in raw sewage, treated wastewater, seawater exposed to wastewater, and fecal samples and/or intestinal homogenates from a wide variety of animals. PMMoV was present in all wastewater samples at concentrations greater than 1 million copies per milliliter of raw sewage.

Though there have been no major outbreaks in humans, the transmittance of this plant virus to humans has created a discussion on the viability of humans as vectors for plant viruses.

Nematodes

Soil-borne nematodes also have been shown to transmit viruses. They acquire and transmit them by feeding on infected roots. Viruses can be transmitted both non-persistently and persistently, but there is no evidence of viruses being able to replicate in nematodes. The virions attach to the stylet (feeding organ) or to the gut when they feed on an infected plant and can then detach during later feeding to infect other plants. Examples of viruses that can be transmitted by nematodes include tobacco ringspot virus and tobacco rattle virus.

Plasmodiophorids

A number of virus genera are transmitted, both persistently and non-persistently, by soil borne zoosporic protozoa. These protozoa are not phytopathogenic themselves, but parasitic. Transmission of the virus takes place when they become associated with the plant roots. Examples include *Polymyxa graminis*, which has been shown to transmit plant viral diseases in cereal crops and *Polymyxa betae* which transmits Beet necrotic yellow vein virus. Plasmodiophorids also create wounds in the plant›s root through which other viruses can enter.

Seed and Pollen Borne Viruses

Plant virus transmission from generation to generation occurs in about 20% of plant viruses. When viruses are transmitted by seeds, the seed is infected in the generative cells and the virus is maintained in the germ cells and sometimes, but less often, in the seed coat. When the growth and development of plants is delayed because of situations like unfavourable weather, there is an increase in the amount of virus infections in seeds. There does not seem to be a correlation between the location of the seed on the plant and its chances of being infected. Little is known about the mechanisms involved in the transmission of plant viruses via seeds, although it is known that it is environmentally influenced and that seed transmission occurs because of a direct invasion of the embryo via the ovule or by an indirect route with an attack on the embryo mediated by infected gametes. These processes can occur concurrently or separately depending on the host plant. It is unknown how the virus is able to directly invade and cross the embryo and boundary between the parental and progeny generations in the ovule. Many plants species can be infected through seeds including but not limited to the families Leguminosae, Solanaceae, Compositae, Rosaceae, Cucurbitaceae, Gramineae. Bean common mosaic virus is transmitted through seeds.

Direct Plant-to-Human Transmission

Researchers from the University of the Mediterranean in Marseille, France have found tenuous evidence that suggest a virus common to peppers, the Pepper Mild Mottle Virus (PMMoV) may have moved on to infect humans. This is a very rare and highly unlikely event as, to enter a cell and replicate, a virus must "bind to a receptor on its surface, and a plant virus would be highly unlikely to recognize a receptor on a human cell. One possibility is that the virus does not infect human cells directly. Instead, the naked viral RNA may alter the function of the cells through a mechanism similar to RNA interference, in which the presence of certain RNA sequences can turn genes on and off," according to Virologist Robert Garry from the Tulane University in New Orleans, Louisiana.

Translation of Plant Viral Proteins

75% of plant viruses have genomes that consist of single stranded RNA (ssRNA). 65% of plant viruses have +ssRNA, meaning that they are in the same sense orientation as messenger RNA but 10% have -ssRNA, meaning they must be converted to +ssRNA before they can be translated. 5% are double stranded RNA and so can be immediately translated as +ssRNA viruses. 3% require a reverse transcriptase enzyme to convert between RNA and DNA. 17% of plant viruses are ssDNA and very few are dsDNA, in contrast a quarter of animal viruses are dsDNA and three quarters of bacteriophage are dsDNA. Viruses use the plant ribosomes to produce the 4-10 proteins encoded by their genome. However, since many of the proteins are encoded on a single strand (that is, they are polycistronic) this will mean that the ribosome will either only produce one protein, as it will terminate translation at the first stop codon, or that a polyprotein will be produced. Plant viruses have had to evolve special techniques to allow the production of viral proteins by plant cells.

5' Cap

For translation to occur, eukaryotic mRNAs require a 5' Cap structure. This means that viruses must also have one. This normally consists of 7MeGpppN where N is normally adenine or guanine. The viruses encode a protein, normally a replicase, with a methyltransferase activity to allow this.

Some viruses are cap-snatchers. During this process, a ^{7m}G-capped host mRNA is recruited by the viral transcriptase complex and subsequently cleaved by a virally encoded endonuclease. The resulting capped leader RNA is used to prime transcription on the viral genome.

However some plant viruses do not use cap, yet translate efficiently due to cap-independent translation enhancers present in 5' and 3' untranslated regions of viral mRNA.

Readthrough

Some viruses (e.g. tobacco mosaic virus (TMV)) have RNA sequences that contain a "leaky" stop codon. In TMV 95% of the time the host ribosome will terminate the synthesis of the polypeptide at this codon but the rest of the time it continues past it. This means that 5% of the proteins produced are larger than and different from the others normally produced, which is a form of translational regulation. In TMV, this extra sequence of polypeptide is an RNA polymerase that replicates its genome.

Production of Sub-Genomic RNAS

Some viruses use the production of subgenomic RNAs to ensure the translation of all proteins within their genomes. In this process the first protein encoded on the genome, and this the first to be translated, is a replicase. This protein will act on the rest of the genome producing negative strand sub-genomic RNAs then act upon these to form positive strand sub-genomic RNAs that are essentially mRNAs ready for translation.

Segmented Genomes

Some viral families, such as the *Bromoviridae* instead opt to have multipartite genomes, genomes split between multiple viral particles. For infection to occur, the plant must be infected with all particles across the genome. For instance *Brome mosaic virus* has a genome split between 3 viral particles, and all 3 particles with the different RNAs are required for infection to take place.

Polyprotein Processing

This strategy is adopted by viral genera such as the Potyviridae and Tymoviridae. The ribosome translates a single protein from the viral genome. Within the polyprotein is an enzyme (or enzymes) with proteinase function that is able to cleave the polyprotein into the various single proteins or just cleave away the protease, which can then cleave other polypeptides producing the mature proteins.

Well understood plant viruses

Tobacco mosaic virus (TMV) and Cauliflower mosaic virus (CaMV) are frequently used in plant molecular biology. Of special interest is the CaMV 35S promoter, which is a very strong promoter most frequently used in plant transformations.

REFERENCES

1. Roossinck, M. J. (2011). "The good viruses: viral mutualistic symbioses". Nature Reviews Microbiology. 9: 99–108. doi:10.1038/nrmicro2491.
2. Oparka KJ and Alison GR, Plasmodesmata. A Not So Open-and-Shut Case. Plant Physiology, Jan. 2001, Vol. 125, pp 123-126
3. Alberts B, Johnson A, Lewis J, Raff M, Roberts K, Walter P (2002). Molecular Biology of the Cell. Garland Science. ISBN 0-8153-3218-1. Chapter 7: Control of Gene Expression, pp 451–452.
4. Ding, S. W.; Voinnet, O. (2007). "Antiviral Immunity Directed by Small RNAs". Cell. 130 (3): 413–426. doi:10.1016/j.cell.2007.07.039. PMC 2703654 . PMID 17693253. PMID 17693253
5. Virus Structure
6. Gray, Stewart M.; Banerjee, Nanditta. "Mechanisms of Arthropod Transmission of Plant and Animal Viruses". Microbiology and Molecular Biology Reviews. 63: 128–148.
7. Verchot-Lubicz, Jeanmarie, "Soilborne viruses: advances in virus movement, virus induced gene silencing, and engineered resistance", Physiological and Molecular Plant Pathology 62, 2003, p. 56
8. Kanyuka, Konstantin; Ward, Elaine; Adams, Michael J. "Polymyxa graminis and the cereal viruses it transmits: a research challenge". Molecular Plant Pathology. 4: 393–406. doi:10.1046/j.1364-3703.2003.00177.x.
9. Colson P, Richet H, Desnues C, Balique F, Moal V, et al. (2010). Mylonakis E, ed. "Pepper Mild Mottle Virus, a Plant Virus Associated with Specific Immune Responses, Fever, Abdominal Pains, and Pruritus in Humans". PLoS ONE. PLoS ONE. 5 (4): e10041. doi:10.1371/journal.pone.0010041. PMC 2850318 . PMID 20386604.
10. "Evidence of First Virus That Moves from Plants to Humans". TechVert. 2010-04-15.
11. Hull, Robert (November 2001). "Classifying reverse transcribing elements: a proposal and a challenge to the ICTV". Archives of Virology. Springer Wien. 146 (11): 2255–2261. doi:10.1007/s007050170036. PMID 11765927. Retrieved 4 July 2009.
12. Duijsings; et al. "In vivo analysis of the TSWV cap-snatching mechanism: single base complementarity and primer length requirements". The EMBO Journal. 20: 2545–2552. doi:10.1093/emboj/20.10.2545.
13. Kneller; et al. (2006). "Cap-independent translation of plant viral RNAs". Virus Research. 119 (1): 63–75. doi:10.1016/j.virusres.2005.10.010.
14. "PP-212/PP134: Tospoviruses (Family Bunyaviridae, Genus Tospovirus)". Edis. ifas.ufl.edu. Retrieved 2014-06-15.
15. Jones, D. R. 2005. Plant viruses transmitted by thrips. European Journal of Plant Pathology 113: 119–157.

16. Mound, L. A. 2002. So many thrips – so few tospoviruses?, pp. 15 - 18. In L. A. Mound and R. Marullo [eds.], Thrips and Tospoviruses: Proceedings of the 7th International Symposium on Thysanoptera. CSIRO Entomology, Reggio Calabria, Italy.
17. Moritz, G., S. Kumm, and L. A. Mound. 2004. Tospovirus transmission depends on thrips ontogeny. Virus research 100: 143–149.
18. Groves, R. L., J. F. Walgenbach, J. W. Moyer, and G. G. Kennedy. 2003. Seasonal Dispersal Patterns of Frankliniella fusca (Thysanoptera: Thripidae) and Tomato Spotted Wilt Virus Occurrence in Central and Eastern North Carolina. Journal of Economic Entomology 96: 1 - 11.
19. Nault, B. A., J. Speese III, D. Jolly, and R. L. Groves. 2003. Seasonal patterns of adult thrips dispersal and implications for management in eastern Virginia tomato fields. Crop Protection 22: 505–512.
20. http://www.oznet.ksu.edu/tospovirus/hostlist.html
21. Materials and Methods for the Detection of Viral Inclusions
22. Inclusions of Tomato spotted wilt virus and Impatiens necrotic spot
23. Whitfield AE, Ullman DE, German TL (2005). Tospovirus-thrips interactions. Annual Review of Phytopathology. 43: 459–89.
24. Hu Q, Hollunder J, Niehl A, Kørner CJ, Gereige D, Windels D, Arnold A, Kuiper M, Vazquez F, Pooggin M, Heinlein M (2011). "Specific impact of tobamovirus infection on the Arabidopsis small RNA profile". PLOS ONE. 6 (5): e19549. doi:10.1371/journal.pone.0019549. PMC 3091872 . PMID 21572953.
25. Plant Pathology Fact Sheet: Pepper mild mottle virus http://ufdc.ufl.edu/IR00001690/00001
26. Mnari-Hattab, M.; Ezzaier, K. (2006). "Biological, Serological, and Molecular Characterization of Pepper mild mottle virus (PMMoV) in Tunisia" (PDF). Tunisian Journal of Plant Protection. 1 (1): 1–12.
27. Jarret, R.L.; Gillaspie, A.G.; Barkley, N.A.; Pinnow, D.L. (2008). "The Occurrence and Control of Pepper Mild Mottle Virus (PMMoV) in the USDA/ARS Capsicum Germplasm Collection". Seed Technology Journal. 30 (1): 26–36.
28. http://edis.ifas.ufl.edu/cv275Archived August 6, 2010, at the Wayback Machine.
29. Wetter, C. (1984). "Serological Identification of Four Tobamoviruses Infected Pepper" (PDF). Plant Disease. 68 (7): 597–599. doi:10.1094/pd-69-597.
30. "Peppers, Tomatoes, and Tobamoviruses" Plant Pathology Circular No. 400 (2000) http://www.freshfromflorida.com/content/download/11406/144732/pp400.pdf
31. Nagvi, S.A.M.H. (ed.). Diseases of Fruits and Vegetables: Diagnosis and Management. Springer. ISBN 9781402018268.
32. "Pepper Mild Mottle Virus." American Vegetable Grower, 59.9 (2011): 30.
33. Yoshimoto, Ryota; Sasaki, Hirotaka; Takahashi, Tadashi; Kanno, Hitoshi; Nanzyo, Masami (2012). "Contribution of Soil Components to Adsorption of Pepper Mild

Mottle Virus by Japanese Soils". Soil Biology & Biochemistry. 46: 96–102. doi:10.1016/j.soilbio.2011.12.006.

34. http://ufdc.ufl.edu/IR00001690/0000

35. Colson, P; Richet, H; Desnues, C; Balique, F; Moal, V; et al. (2010). "Pepper Mild Mottle Virus, a Plant Virus Associated with Specific Immune Responses, Fever, Abdominal Pains, and Pruritus in Humans". PLoS ONE. 5 (4): e10041. doi:10.1371/journal.pone.0010041. PMC 2850318 . PMID 20386604.

36. Rosario, K.; Symonds, E.M.; Sinigalliano, C.; Stewart, J.; Breitbart, M. (2009). "Pepper Mild Mottle Virus as an Indicator of Fecal Pollution"(PDF). Applied and Environmental Microbiology. 75 (22): 7261–7267. doi:10.1128/AEM.00410-09.

CHAPTER 4

Physiological Plant Disorder

Chapter Contents

Physiological disorders of plants are those problems resulting from the influence of environmental and cultural factors on plant development. A number of physiological disorders arise in controlled environment facilities because environmental and cultural conditions in these facilities often are significantly different from those encountered by plants in the natural environment.

INTRODUCTION

Physiological plant disorders are caused by non-pathological conditions such as poor light, adverse weather, water-logging, phytotoxic compounds or a lack of nutrients, and affect the functioning of the plant system. Physiological disorders are distinguished from plant diseases caused by pathogens, such as a virus or fungus.While the symptoms of physiological disorders may appear disease-like, they can usually be prevented by altering environmental conditions. However, once a plant shows symptoms of a physiological disorder it is likely that that season's growth or yield will be reduced.

Diagnosis of the cause of a physiological disorder (or disease) can be difficult, but there are many web-based guides that may assist with this. Examples are: *Abiotic plant disorders: Symptoms, signs and solutions*; *Georgia Corn Diagnostic Guide*;*Diagnosing Plant Problems* (Kentucky); and *Diagnosing Plant Problems* (Virginia).

Sunburn at apple.

Some general tips to diagnosing plant disorders:

- Examine where symptoms first appear on a plant—on new leaves, old leaves or all over?
- Note the pattern of any discolouration or yellowing—is it all over, between the veins or around the edges? If only the veins are yellow deficiency is probably not involved.
- Note general patterns rather than looking at individual plants—are the symptoms distributed throughout a group of plants of the same type growing together. In the case of a deficiency all of the plants should be similarly effected, although distribution will depend on past treatments applied to the soil.
- Soil analysis, such as determining pH, can help to confirm the presence of physiological disorders.
- Consider recent conditions, such as heavy rains, dry spells, frosts, etc., may also help to determine the cause of plant disorders.

Deficit of micronutriens, vine.

Weather Damage

Frost and cold are major causes of crop damage to tender plants, although hardy plants can also suffer if new growth is exposed to a hard frost following a period of warm weather. Symptoms will often appear overnight, affecting many types of plants. Leaves and stems may turn black, and buds and flowers may be discoloured, and frosted blooms may not produce fruit. Many annual plants, or plants grown in frost free areas, can suffer from damage when the air temperature drops below 40 degrees Fahrenheit (4 degrees Celsius). Tropical plants may begin to experience cold damage when the temperature is 42 to 48 °F (5 to 9 °C), symptoms include wilting of the top of the stems and/or leaves, and blackening or softening of the plant tissue.

Frost or cold damage can be avoided by ensuring that tender plants are properly hardened before planting, and that they are not planted too early in the season, before the risk of frost has passed. Avoid planting susceptible plants in frost pockets, or where they will receive early morning sun. Protect young buds and bloom with horticultural fleece if frost is forecast. Cold, drying easterly winds can also severely inhibit spring growth even without an actual frost, thus adequate shelter or the use of windbreaks is important.

Drought can cause plants to suffer from water stress and wilt. Adequate irrigation is required during prolonged hot, dry periods. Rather than shallow daily watering, during a drought water should be directed towards the roots, ensuring that the soil is thoroughly soaked two or three times a week. Mulches also help preserve soil moisture and keep roots cool.

Heavy rains, particularly after prolonged dry periods, can also cause roots to split, onion saddleback (splitting at the base), tomatoes split and potatoes to become deformed or hollow. Using mulches or adding organic matter such as leaf mold, compost or well-rotted manure to the soil will help to act as a 'buffer' between sudden changes in conditions. Water-logging can occur on poorly drained soils, particularly following heavy rains. Plants can become yellow and stunted, and will tend to be more prone to drought and diseases. Improving drainage will help to alleviate this problem.

Hail can cause damage to soft skinned fruits, and may also allow brown rot or other fungi to penetrate the plant. Brown spot markings or lines on one side of a mature apple are indicative of a spring hailstorm.

Plants affected by salt stress are unable to take water from soil, due to an osmotic imbalance between soil and plant.

Iron deficiency.

Nutrient Deficiencies

Poor growth and a variety of disorders such as leaf discolouration (chlorosis) can be caused by a shortage of one or more plant nutrients. Poor plant uptake of a nutrient from the soil (or other growing medium) may be due to an absolute shortage of that element in the growing medium, or because that element is present in a form that is not available to the plant. The latter can be caused by incorrect pH, shortage of water, poor root growth or an excess of another nutrient. Plant nutrient deficiencies can be avoided or corrected using a variety of approaches including the consultation of experts on-site, the use of soil and plant-tissue testing services, the application of prescription-blend fertilizers, the application of fresh or well-decomposed organic matter, and the use of biological systems such as cover crops, intercropping, improved fallows, ley cropping, permaculture, or crop rotation.

Drought.

Nutrient (or mineral) deficiencies include:

- Boron deficiency
- Calcium deficiency
- Iron deficiency
- Magnesium deficiency
- Manganese deficiency
- Nitrogen deficiency
- Phosphorus deficiency
- Potassium deficiency

- Zinc deficiency
- Shortage of trace elements such as molybdenum can also cause disorders such as whiptail in cauliflower.

BORON DEFICIENCY (PLANT DISORDER)

Boron deficiency is a common deficiency of the micronutrient boron in plants. It is the most widespread micronutrient deficiency around the world and causes large losses in crop production and crop quality. Boron deficiency affects vegetative and reproductive growth of plants, resulting in inhibition of cell expansion, death of meristem, and reduced fertility.

Plants contain boron both in a water-soluble and insoluble form. In intact plants, the amount of water-soluble boron fluctuates with the amount of boron supplied, while insoluble boron does not. The appearance of boron deficiency coincides with the decrease of water-insoluble boron. It appears that the insoluble boron is the functional form while the soluble boron represents the surplus.

Boron is essential for the growth of higher plants. The primary function of the element is to provide structural integrity to the cell wall in plants. Other functions likely include the maintenance of the plasma membrane and other metabolic pathways.

Symptoms

Symptoms include dying growing tips and bushy stunted growth, extreme cases may prevent fruit set. Crop-specific symptoms include;

- *Apple*- interacting with calcium, may display as “water core”, internal areas appearing frozen
- *Beetroot*- rough, cankered patches on roots, internal brown rot.
- *Cabbage*- distorted leaves, hollow areas in stems.
- *Cauliflower*- poor development of curds, and brown patches. Stems, leafstalks and midribs roughened.
- *Celery*- leaf stalks develop cracks on the upper surface, inner tissue is reddish brown.
- *Celeriac*- causes brown heart rot
- *Pears*- new shoots die back in spring, fruits develop hard brown flecks in the skin.
- *Strawberries*- Stunted growth, foliage small, yellow and puckered at tips. Fruits are small and pale.
- *Swede (rutabaga)* and *turnip*- brown or gray concentric rings develop inside the roots.
- *Arecaceae* (*Palm Tree*) - brown spots on fronds & lower productivity.

Soil Conditions

Boron is present in the soil in many forms, the most common being Boric Acid (H_3BO_3). An adequate amount of boron in the soil is 12 mg/kg. If the boron content of the soil drops below 0.14 mg/kg then boron deficiency is likely to be observed. Boron deficiency is also observed in basic soils with a high pH because in basic conditions boric acid exists in an undissociated form which the plant is unable to absorb. Soils with low organic matter content (<1.5%) are also susceptible to boron deficiency. Highly leached sandy soils are also characteristic of boron deficiency because the boron will not be retained in the soil.Boron toxicity is also possible if the boron content of the soil is high enough that the plant cannot cope with the excess boron. The levels at which boron is toxic to plants varies with different species of plants.

Boron Requirements

Boron is an essential micronutrient which means it is essential for plant growth and development, but is required in very small quantities. Although Boron requirements vary among crops, the optimum boron content of the leaves for most crops is 20-100 ppm. Excess boron can result in boron toxicity and the toxicity level varies between plants.

Treatment

Boric acid (16.5%boron), borax (11.3% boron) or SoluBor (20.5% boron) can be applied to soils to correct boron deficiency. Typical applications of actual boron are about 1.1 kg/hectare or 1.0 lb/acre but optimum levels of boron vary with plant type.Borax, Boric Acid or Solubor can be dissolved in water and sprayed or applied to soil as a dust. Excess boron is toxic to plants so care must be taken to ensure correct application rate and even coverage. Leaves of many plants are damaged by boron; therefore, when in doubt, only apply to soil. Application of boron may not correct boron deficiency in alkaline soils because even with the addition of boron, it may remain unavailable for plant absorption. Continued application of boron may be necessary in soils that are susceptible to leaching such as sandy soils. Flushing soils containing toxic levels of boron with water can remove the boron through leaching.

Functions

Once boron has been absorbed by the plant and incorporated into the various structures that require boron, it is unable to disassemble these structures and re-transport boron through the plant resulting in boron being a non-mobile nutrient. Due to translocation difficulties the youngest leaves often show deficiency symptoms first.

Cell Wall

Boron is part of the dRG-II-B complex which is involved in the cross linking for pectin located in the primary cell wall and the middle lamella of plant cells. This cross linking is thought to stabilize the matrix of plant cell walls.

Causes Change in Carbohydrate Metabolism

Enhances Protein Synthesis

Germination and Pollination

The B requirement is much higher for reproductive growth than for vegetative growth in most plant species. Boron increases flower production and retention, pollen tube elongation and germination, and seed and fruit development.

A deficiency of B can cause incomplete pollination of corn or prevent maximum pod-set in soybeans.

Sugar Translocation

Photosynthesis transforms sunlight energy into plant energy compounds such as sugars. For this process to continue in plants, the sugars must be moved away from the site of their development, and stored or used to make other compounds.

Boron increases the rate of transport of sugars (which are produced by photosynthesis in mature plant leaves) to actively growing regions and also in developing fruits. Boron is essential for providing sugars which are needed for root growth in all plants and also for normal development of root nodules in legumes such as alfalfa, soybeans and peanuts.

CALCIUM DEFICIENCY (PLANT DISORDER)

Calcium (Ca) deficiency is a plant disorder that can be caused by insufficient calcium in the growing medium, but is more frequently a product of low transpiration of the whole plant or more commonly the affected tissue. Plants are susceptible to such localized calcium deficiencies in low or non-transpiring tissues because calcium is not transported in the phloem. This may be due to water shortages, which slow the transportation of calcium to the plant, poor uptake of calcium through the stem, or too much nitrogen in the soil.

Calcium deficiency (blossom end rot) on a tomato

Causes

Acidic, sandy, or coarse soils often contain less calcium. Uneven soil moisture and overuse of fertilizers can also cause calcium deficiency. At times, even with sufficient calcium in the soil, it can be in an insoluble form and is then unusable by the plant or it could be attributed to a "transport protein". Soils containing high phosphorus are particularly susceptible to creating insoluble forms of calcium.

Symptoms

Blossom end rot on a grape tomato

Calcium deficiency symptoms appear initially as localized tissue necrosis leading to stunted plant growth, necrotic leaf margins on young leaves or curling of the leaves, and eventual death of terminal buds and root tips. Generally, the new growth and rapidly growing tissues of the plant are affected first. The mature leaves are rarely if ever affected because calcium accumulates to high concentrations in older leaves.

Crop-specific symptoms include:

Apple

'Bitter pit' – fruit skins develop pits, brown spots appear on skin and/or in flesh and taste of those areas is bitter. This usually occurs when fruit is in storage, and Bramley apples are particularly susceptible. Related to boron deficiency, "water cored" apples seldom display bitter pit effects.

Cabbage and Brussels sprouts

Internal browning and "tip burn"

Carrot

'Cavity spot' – oval spots develop into craters which may be invaded by other disease-causing organisms.

Celery

Stunted growth, central leaves stunted.

Tomatoes and peppers

'Blossom end rot' – Symptoms start as sunken, dry decaying areas at the blossom end of the fruit, furthest away from the stem, not all fruit on a truss is necessarily affected. Sometimes rapid growth from high-nitrogen fertilizers may exacerbate blossom end rot. Water management and preventing water stress is key to minimizing its occurrence.

Treatment

Calcium deficiency can sometimes be rectified by adding agricultural lime to acid soils, aiming at a pH of 6.5, unless the subject plants specifically prefer acidic soil. Organic matter should be added to the soil to improve its moisture-retaining capacity. However, because of the nature of the disorder (i.e. poor transport of calcium to low transpiring tissues), the problem cannot generally be cured by the addition of calcium to the roots. In some species, the problem can be reduced by prophylactic spraying with calcium chloride of tissues at risk.

Dissection of grape tomato with blossom end rot

Plant damage is difficult to reverse, so corrective action should be taken immediately, supplemental applications of calcium nitrate at 200 ppm nitrogen, for example. Soil pH should be tested, and corrected if needed, because calcium deficiency is often associated with low pH.

Early fruit will generally have the worst systems, with them typically lessening as the season progresses. Preventative measures, such as irrigating prior to especially high temperatures and stable irrigation will minimize the occurrence.

IRON DEFICIENCY (PLANT DISORDER)

Iron (Fe) deficiency is a plant disorder also known as "lime-induced chlorosis". It can be confused with manganese deficiency. A deficiency in the soil is rare but iron can be unavailable for absorption if soil pH is not between about 5 and 6.5. A common problem is excessive alkalinity of the soil (the pH is above 6.5). Also, iron deficiency can develop if the soil is too waterlogged or has been overfertilised. Elements like calcium, zinc, manganese, phosphorus, or copper can tie up iron if they are present in high amounts.

Iron deficiency showing chlorotic leaves in a lemon tree. Compare yellow clorotic leaves with the green non chlorotic leaves at left of this image

Iron is needed to produce chlorophyll, hence its deficiency causes chlorosis. For example, iron is used in the active site of glutamyl-tRNA reductase, an enzyme needed for the formation of 5-Aminolevulinic acid which is a precursor of heme and chlorophyll.

Symptoms

Symptoms include leaves turning yellow or brown in the margins between the veins which may remain green, while young leaves may appear to be bleached. Fruit would be of poor quality and quantity. Any plant may be affected, but raspberries and pears are particularly susceptible, as well as most acid-loving plants such as azaleas and camellias.

Treatment

Iron deficiency can be avoided by choosing appropriate soil for the growing conditions (e.g., avoid growing acid loving plants on lime soils), or by adding well-rotted manure or compost. If iron deficit chlorosis is suspected then check the pH of the soil with an appropriate test kit or instrument. Take a soil sample at surface and at depth. If the pH is over seven then consider soil remediation that will lower the pH toward the 6.5 - 7 range. Remediation includes: i) adding compost, manure, peat or similar organic matter (warning. Some retail blends of manure and compost have pH in the range 7 - 8 because of added lime. Read the MSDS if available. Beware of herbicide residues in manure. Source manure from a certified organic source.) ii) applying Ammonium Sulphate as a Nitrogen fertilizer (acidifying fertilizer due to decomposition of ammonium ion to nitrate in the soil and root zone) iii) applying elemental Sulphur to the soil (oxidizes over the course of months to produce sulphate/sulphite and lower pH). Note: adding acid directly e.g. sulphuric/hydrochloric/citric acid is dangerous as you may mobilize metal ions in the soil that are toxic and otherwise bound. Iron can be made available immediately to the plant by the use of iron sulphate or iron chelate compounds. Two common iron chelates are Fe EDTA and Fe EDDHA. Iron sulphate (Iron(II)_sulfate) and iron EDTA are only useful in soil up to PH 7.1 but they can be used as a foliar spray (Foliar_feeding). Iron EDDHA is useful up to PH 9 (highly alkaline) but must be applied to the soil and in the evening to avoid photodegradation. EDTA in the soil may mobilize Lead, EDDHA does not appear to.

MAGNESIUM DEFICIENCY (PLANTS)

Magnesium (Mg) deficiency is a detrimental plant disorder that occurs most often in strongly acidic, light, sandy soils, where magnesium can be easily leached away. Magnesium is an essential micro nutrient found from 0.2-0.4% dry matter and is necessary for normal plant growth.

A plant with Magnesium deficiency

Excess potassium, generally due to fertilizers, further aggravates the stress from the magnesium deficiency, as does aluminium toxicity.

Magnesium has an important role in photosynthesis because it forms the central atom of chlorophyll. Therefore, without sufficient amounts of magnesium, plants begin to degrade the chlorophyll in the old leaves. This causes the main symptom of magnesium deficiency, chlorosis, or yellowing between leaf veins, which stay green, giving the leaves a marbled appearance. Due to magnesium's mobile nature, the plant will first break down chlorophyll in older leaves and transport the Mg to younger leaves which have greater photosynthetic needs. Therefore, the first sign of magnesium deficiency is the chlorosis of old leaves which progresses to the young leaves as the deficiency continues. Magnesium also is a necessary activator for many critical enzymes, including ribulosbiphosphate carboxylase (RuBisCO) and phosphoenolpyruvate carboxylase (PEPC), both essential enzymes in carbon fixation. Thus low amounts of Mg lead to a decrease in photosynthetic and enzymatic activity within the plants. Magnesium is also crucial in stabilizing ribosome structures, hence, a lack of magnesium causes depolymerization of ribosomes leading to pre-mature aging of the plant. After prolonged magnesium deficiency, necrosis and dropping of older leaves occurs. Plants deficient in magnesium also produce smaller, woodier fruits.

Magnesium deficiency may be confused with zinc or chlorine deficiencies, viruses, or natural ageing since all have similar symptoms. Adding Epsom salts (as a solution of 25 grams per liter or 4 oz per gal) or crushed dolomitic limestone to the soil can rectify magnesium deficiencies. For a more organic solution, applying home-made compost mulch can prevent leaching during excessive rainfall and provide plants with sufficient amounts of nutrients, including magnesium.

MANGANESE DEFICIENCY (PLANT)

Manganese (Mn) deficiency is a plant disorder that is often confused with, and occurs with, iron deficiency. Most common in poorly drained soils, also where organic matter levels are high. Manganese may be unavailable to plants where pH is high.

Affected plants include onion, apple, peas, French beans, cherry and raspberry, and symptoms include yellowing of leaves with smallest leaf veins remaining green to produce a 'chequered' effect. The plant may seem to grow away from the problem so that younger leaves may appear to be unaffected. Brown spots may appear on leaf surfaces, and severely affected leaves turn brown and wither.

Prevention can be achieved by improving soil structure. Do not over-lime.

Manganese deficiency on a rose

NITROGEN DEFICIENCY

All plants require sufficient supplies of macronutrients for healthy growth, and nitrogen (N) is a nutrient that is commonly in limited supply. Nitrogen deficiency in plants can occur when organic matter with high carbon content, such as sawdust, is added to soil. Soil organisms use any nitrogen to break down carbon sources, making N unavailable to plants. This is known as "robbing" the soil of nitrogen. All vegetables apart from nitrogen fixing legumes are prone to this disorder.

Nitrogen deficiency can be prevented in the short term by using grass mowings as a mulch, or foliar feeding with manure, and in the longer term by building up levels of organic matter in the soil. Sowing green manure crops such as grazing rye to cover soil over the winter will help to prevent nitrogen leaching, while leguminous green manures such as winter tares will fix additional nitrogen from the atmosphere.

A young cabbage plant exhibiting nitrogen deficiency.

PHOSPHORUS DEFICIENCY

Phosphorus deficiency is a plant disorder associated with insufficient supply of phosphorus. Phosphorus refers here to salts of phosphates (PO_4^{3-}), monohydrogen phosphate (HPO_4^{2-}), and dihydrogen phosphate ($H_2PO_4^{-}$). These anions readily interconvert, and the predominant species is determined by the pH of the solution or soil. Phosphates are required for the biosynthesis of genetic material as well as ATP, essential for life. Phosphorus deficiency can be controlled by applying sources of phosphorus such as bone meal, rock phosphate, manure, and phosphate-fertilizers.

Phosphorus deficiency on corn

Symptoms (Biological Implications)

In plants, Phosphorus (P) is considered second to nitrogen as the most essential nutrient to ensure health and function. Phosphorus is used by plants in numerous processes such as photophosphorylation, genetic transfer, the transportation of nutrients, and phospholipid cell membranes.Within a plant cell these functions are imperative for function, in photophosphoroylation for example the creation of stored energy in plants is a result of a chemical reaction including phosphorus. Phosphorus is a key molecular component of genetic reproduction. When phosphorus is present in inadequate levels, genetic processes such as cell division and plant growth are impaired. Hence, phosphorus deficient plants may mature at a slower rate than plants with adequate amounts of phosphorus. The stunted growth induced by phosphorus deficiency has been correlated with smaller leaf sizes and a lessened number of leaves. Phosphorus deficiency may also create an imbalance in the storage of carbohydrates. Photosynthesis, the main function of plant cells that produces energy from sunlight and water, usually remains at a normal rate under a phosphorus-deficient state. However phosphorus usage in functions within the cell usually slow. This imbalance of rates in phosphorus deficient plants leads to the buildup of excess carbohydrate within the plant. This carbohydrate buildup often can be observed by the darkening of leaves. In some plants the leaf pigment change as a result of this process can turn leaves a dark purplish color.

Detection

Detecting phosphorus deficiency can take multiple forms. A preliminary detection method is a visual inspection of plants. Darker green leaves and purplish or red pigment can indicate a deficiency in phosphorus. This method however can be an unclear diagnosis because other plant environment factors can result in similar discoloration symptoms. In commercial or well monitored settings for plants, phosphorus deficiency is diagnosed by scientific testing. Additionally, discoloration in plant leaves only occurs under fairly severe phosphorus deficiency so it is beneficial to planters and farmers to scientifically check phosphorus levels before discoloration occurs. The most prominent method of checking phosphorus levels is by soil testing. The major soil testing methods are Bray 1-P, Mehlich 3, and Olsen methods. Each of these methods are viable but each method has tendencies to be more accurate in known geographical areas. These tests use chemical solutions to extract phosphorus from the soil. The extract must then be analyzed to determine the concentration of the phosphorus. Colorimetry is used to determine this concentration. With the addition of the phosphorus extract into a colorimeter, there is visual color change of the solution and the degree to this color change is an indicator of phosphorus concentration. To apply this testing method on phosphorus deficiency, the measured phosphorus concentration must be compared to known values. Most plants have established and thoroughly tested optimal soil conditions. If the concentration of phosphorus measured from the colorimeter test is significantly lower than the plant's optimal soil levels, then it is likely the plant is phosphorus deficient. The soil testing with colorimetric analysis, while widely used, can be subject to diagnostic problems as a result of interference from other present compounds and elements. Additional phosphorus detection methods such as spectral radiance and inductively coupled plasma spectrometry (ICP) are also implemented with the goal of improving reading accuracy. According to the World Congress of Soil Scientists, the advantages of these light-based measurement methods are their quickness of evaluation, simultaneous measurements of plant nutrients, and their non-destructive testing nature. Although these methods have experimental based evidence, unanimous approval of the methods has not yet been achieved.

Treatment

Correction and prevention of phosphorus deficiency typically involves increasing the levels of available phosphorus into the soil. Planters introduce more phosphorus into the soil with bone meal, rock phosphate,manure, and phosphate-fertilizers. The introduction of these compounds into the soil however does not ensure the alleviation of phosphorus deficiency. There must be phosphorus in the soil, but the phosphorus must also be absorbed by the plant. The uptake of phosphorus is limited by the chemical form in which the phosphorus is available in the soil. A large percentage of phosphorus in soil is present in chemical compounds that plants are incapable of absorbing. Phosphorus must be present in soil in specific chemical arrangements to be usable as plant nutrients. Facilitation of usable phosphorus in soil can be optimized by maintaining soil within a specified pH range. Soil acidity, measured on the pH scale, partially dictates what chemical arrangements that phosphorus forms. Between pH 6 and 7, phosphorus makes the fewest number of bonds which render the nutrient unusable to plants. At this range of acidity the likeliness of phosphorus uptake is increased and the likeliness of phosphorus deficiency is decreased. Another component in the prevention and treatment of phosphorus is the plant's disposition to absorb nutrients. Plant species and different plants within in the same species react differently to low levels of phosphorus in soil. Greater expansion of root systems generally correlate to greater nutrient uptake. Plants within a species that have larger roots are genetically advantaged and less prone to phosphorus deficiency. These plants can be cultivated and bred as a long term phosphorus deficiency prevention method. In conjunction to root size, other genetic root adaptations to low phosphorus conditions such as mycorrhizal symbioses have been found to increase nutrient intake. These biological adaptations to roots work to maintain the levels of vital nutrients. In larger commercial agriculture settings, variation of plants to adopt these desirable phosphorus intake adaptations may be a long-term phosphorus deficiency correction method.

POTASSIUM DEFICIENCY (PLANTS)

Potassium deficiency, also known as potash deficiency, is a plant disorder that is most common on light, sandy soils, because potassium ions (K^+) are highly soluble and will easily leach from soils without colloids. Potassium deficiency is also common in chalky or peaty soils with a low clay content. It is also found on heavy clays with a poor structure.

A potassium-deprived leaf

Role of Potassium in Plants

The main role of potassium is to provide the ionic environment for metabolic processes in the cytosol, and as such functions as a regulator of various processes including growth regulation. Plants require potassium ions (K^+) for protein synthesis and for the opening and closing of stomata, which is regulated by proton pumps to make surrounding guard cells either turgid or flaccid. A deficiency of potassium ions can impair a plant's ability to maintain these processes. Potassium also functions in other physiological processes such as photosynthesis, protein synthesis, activation of some enzymes, phloem solute transport of photoassimilates into source organs, and maintenance of cation:anion balance in the cytosol and vacuole.

Symptoms of Potassium Deficiency

Typical symptoms of potassium deficiency in plants include brown scorching and curling of leaf tips as well as chlorosis (yellowing) between leaf veins. Purple spots may also appear on the leaf undersides. Plant growth, root development, and seed and fruit development are usually reduced in potassium-deficient plants. Often, potassium deficiency symptoms first appear on older (lower) leaves because potassium is a mobile nutrient, meaning that a plant can allocate potassium to younger leaves when it is K deficient. Deficient plants may be more prone to frost damage and disease, and their symptoms can often be confused with wind scorch or drought. The deficiency is most common in several important fruit and vegetable crops; notably potatoes, brassicas, tomatoes, apples, currants, gooseberries, and raspberries. Sugar beets, cereals, and clover are also commonly affected. Specific symptoms for each of these plants are as follows:

In potatoes, tuber size is much reduced and crop yield is low. The leaves of the plant appear dull and are often blue-green in color with intervenal chlorosis. Leaves will also develop small, dark brown spots on the undersides and a bronzed appearance on the upper surfaces.

In brassicas, leaves are blue-green in color and may have a low degree of intervenal chlorosis. Scorching along the outside edges of leaves is common, and leaves are often tough in texture due to slow growth.

In tomatoes, the stems are woody and growth is slow. Leaves are blue-green in color, and the intervenal area often fades to a pale gray color. Leaves may also have a bronzed appearance and yellow and orange patches may develop on some of the leaflets. Fruits often ripen unevenly and sometimes have green patches near the stalks.

In apples, leaves are scorched around the edges, and intervenal chlorosis is common. Apple fruits often have a slightly acidic or woody taste.

In gooseberries, currants, and raspberries, dieback of shoots and branches is common and although the plant may produce many blossom buds in the early stages of deficiency, fruit yields turn out low and the fruits are of poor quality.

Potassium Deficiency and Plant Disease

For many species, potassium-deficient plants are more susceptible to frost damage and certain diseases than plants with adequate potassium levels. Increased disease resistance associated with adequate potassium levels indicates that potassium has roles in providing disease resistance, and increasing the potassium levels of deficient plants have been shown to decrease the intensity of many diseases. However, increasing potassium concentration above the optimal level does not provide greater disease resistance. In agriculture, some cultivars are more efficient at K uptake due to genetic variations, and often these plants have increased disease resistance. The mechanisms involved with increased host resistance and potassium include a decreased cell permeability and decreased susceptibility to tissue penetration. Silica, which is accumulated in greater quantities when adequate potassium is present, is incorporated into cell walls, strengthening the epidermal layer which functions as a physical barrier to pathogens. Potassium has also been implicated to have a role in the proper thickening of cell walls.

Prevention and Cure

The most widely used potassium fertilizer is potassium chloride (muriate of potash). Other inorganic potassium fertilizers include potassium nitrate, potassium sulfate, and monopotassium phosphate. Potassium-rich treatments suitable for organic farming include feeding with home-made comfrey liquid, adding seaweed meal, composted bracken, and compost rich in decayed banana peels. Wood ash also has high potassium content. Adequate moisture is necessary for effective potassium uptake; low soil water reduces K uptake by plant roots. Liming acidic soils can increase potassium retention in some soils by reducing leaching; practices that increase soil organic matter can also increase potassium retention.

ZINC DEFICIENCY (PLANT DISORDER)

Zinc deficiency occurs when plant growth is limited because the plant cannot take up sufficient quantities of this essential micronutrient from its growing medium. It is one of the most widespread micronutrient deficiencies in crops and pastures worldwide and causes large losses in crop production and crop quality.

Almost half of the world's cereal crops are grown on zinc-deficient soils; as a result, zinc deficiency in humans is a widespread problem.

Maize plants with severe zinc deficiency in the foreground, with healthier plants (planted at the same time) in the background.

Symptoms

Macadamia shoots showing zinc deficiency symptoms. The youngest leaves show chlorosis (yellowing), dwarfing and malformation.

Visible Deficiency Symptoms Include:

- Chlorosis - yellowing of leaves; often interveinal; in some species, young leaves are the most affected,but in others both old and new leaves are chlorotic;
- Necrotic spots - death of leaf tissue on areas of chlorosis;
- Bronzing of leaves - chlorotic areas may turn bronze coloured;
- Rosetting of leaves - zinc-deficient dicotyledons often have shortened internodes, so leaves are clustered on the stem;
- Stunting of plants - small plants may occur as a result of reduced growth or because of reduced internode elongation;
- Dwarf leaves (‹little leaf›) - small leaves that often show chlorosis, necrotic spots or bronzing;
- Malformed leaves - leaves are often narrower or have wavy margins.

Soil Conditions

Zinc deficiency is common in many different types of soil; some soils (sandy soils, histosols and soils developed from highly weathered parent material) have low total zinc concentrations, and others have low plant-available zinc due to strong zinc sorption (calcareous soils, highly weathered soils, vertisols, hydromorphic soils, saline soils). Soils low in organic matter (such as where topsoils have been removed), and compacted soils that restrict root proliferation also have a high risk of zinc deficiency. Application of phosphorus fertilizers has frequently been associated with zinc deficiency; this may be due to enhanced sorption by clay minerals (especially iron oxides), suppression of vesicular arbuscular mycorrhizae and/or immobilization of zinc in plant tissues. Liming of soils also frequently induces zinc deficiency by increasing zinc sorption.

Zinc Requirements

Zinc is an essential micronutrient which means it is essential for plant growth and development, but is required in very small quantities. Although zinc requirements vary among crops, zinc leaf concentrations (on a dry matter basis) in the range 20 to 100 mg/kg are adequate for most crops.

Treatment

Zinc sulphate or zinc oxide can be applied to soils to correct zinc deficiency. Recommended applications of actual zinc range from 5 to 100 kg/hectare but optimum levels of zinc vary with plant type and the severity of the deficiency. Application of zinc may not correct zinc deficiency in alkaline soils because even with the addition of zinc, it may remain unavailable for plant absorption.

Foliar applications of zinc as zinc sulphate or as zinc chelate (or other organic complexes) are also widely used, especially with fruit trees and grape vines. Zinc can also be supplied as a seed treatment, or by root-dipping of transplant seedlings.

Functions

Zinc occurs in plants as a free ion, as a complex with a variety of low molecular weight compounds, or as a component of proteins and other macromolecules. In many enzymes, zinc acts as a functional, structural, or regulatory cofactor; a large number of zinc-deficiency disorders are associated with the disruption of normal enzyme activity (including that of key photosynthetic enzymes). Zinc deficiency also increases membrane leakiness as zinc-containing enzymes are involved in the detoxification of membrane-damaging oxygen radicals. Zinc may be involved in the control of gene expression; it appears important in stabilizing RNA and DNA structure, in maintaining the activity of DNA-synthesizing enzymes and in controlling the activity of RNA-degrading enzymes.

REFERENCES

1. Schutzki, R.E.; Cregg, B. (2007). "Abiotic plant disorders: Symptoms, signs and solutions. A diagnostic guide to problem solving" (PDF). http://www.hrt.msu.edu/. Michigan State University. Retrieved 10 April 2015. External link in |website= (help)
2. Lee, R.D. (2012). "Georgia Corn Diagnostic Guide". extension.uga.edu. University of Georgia Cooperative Extension. Retrieved 10 April 2015.
3. Green, J.L.; Maloy, O.; Capizzi, J.; Hartman, J.; Townsend, L. (2011). "Diagnosing Plant Problems: Kentucky Master Gardener Manual Chapter 7" (PDF). www2.

ca.uky.edu. Kentucky Cooperative Extension. Retrieved 10 April 2015.

4. Niemiera, A.X. (2009). "Diagnosing Plant Problems". pubs.ext.vt.edu. Virginia Cooperative Extension. Retrieved 10 April 2015.
5. Mengel, K.; Kirkby, E.A. (2001). Principles of plant nutrition (5th ed.). Dordrecht: Kluwer Academic Publishers. ISBN 079237150X.
6. "Improved fallows". teca.fao.org. FAO. Retrieved 10 April 2015. Improved fallow is also land resting from cultivation but the vegetation comprises planted and managed species of
7. Shorrocks VM (1997). "The occurrence and correction of boron deficiency.". Plant and Soil. 193 (1): 121–148. doi:10.1023/A:1004216126069.
8. Marschner H (1995). Mineral Nutrition of Higher Plants (2nd ed.). San Diego: Academic Press. pp. 379–396.
9. "Boron deficiency". Retrieved 2012-11-21.
10. Camacho-Cristóbal, Juan J.; Jesús Rexach; Agustín González-Fontes. "Boron in plants: deficiency and toxicity" (PDF). Journal of Integrative Plant Science. Retrieved 2012-11-21.
11. Boron the Overlooked Essential Element
12. d www.agnet.org
13. Boron Toxicity Archived July 15, 2011, at the Wayback Machine.
14. Plant Analysis Handbook for Georgia
15. Toshiro Matsunaga, D.Agr. Archived July 20, 2011, at the Wayback Machine.
16. http://www.borax.com/market/agriculture.aspx
17. "Plant Physiology Online: 5.1 Symptoms of Deficiency In Essential Minerals".
18. University of Zurich (2011). Blossom end rot: Transport protein identified. http://phys.org/news/2011-11-blossom-protein.html
19. "Blossom End Rot: How to Identify, Treat, and Prevent It". TomatoRot.com. Retrieved 21 September 2015.
20. E. W. Simon. "The Symptoms of Calcium Deficiency in Plants". pp. 2–4.
21. Dr. Harry A. Mills. "The University of Georgia College of Agricultural and Environmental Sciences, Department of Horticulture - Carrot". pp. 3–4.
22. Fake, Cindy (March 2010). "MANAGING BLOSSOM-END ROT IN TOMATOES AND PEPPERS" (PDF). Horticulture and Small Farms Advisor, Nevada & Placer Counties. UCANR.
23. Schuster, James. "Focus on Plant Problems - Chlorosis". University of Illinois at Urbana-Champaign. Retrieved 2008-12-22.
24. A. Madan Kumar and Dieter Söll, Antisense HEMA1 RNA Expression Inhibits Heme and Chlorophyll Biosynthesis in Arabidopsis, Plant Physiol, January 2000, Vol. 122, pp. 49-56 http://www.plantphysiol.org/cgi/content/full/122/1/49
25. Norman P.A. Huner; William Hopkins. "3 & 4". Introduction to Plant Physiology 4th Edition. John Wiley & Sons, Inc. ISBN 978-0-470-24766-2.

26. Ding Y.; Chang C.; Luo W. "High Potassium Aggravates the Oxidative Stress Induced by Magnesium Deficiency in Rice Leaves". Pedosphere. 18 (3). pp. 316–327.
27. Merhaut, D.J. (2006). "Magnesium". In Barker A.V.; Pilbeam D.J. Handbook of plant nutrition. Boca Raton: CRC Press. p. 154. ISBN 9780824759049.
28. Hermans C.; Vuylsteke F.; Coppens F. "Systems Analysis of the responses to long-term magnesium deficiency and restoration in Arabidopsis thaliana". New Phytologist. 187. pp. 132–144.
29. "Problem Solving: Magnesium Deficiency".
30. http://whatcom.wsu.edu/ag/compost/fundamentals/needs_carbon_nitrogen.htm
31. Pandey, S N; Sinha, B K. "Mineral Nutrition". Plant Physiology (fourth ed.). 576Masjid Road, Jangpura, New Delhi-110014: VIKAS PUBLISHING HOUSE Pvt. Ltd. pp. 125–126. ISBN 8125918795.
32. "Idaho Nutrient Management - Potato". www.extension.uidaho.edu. Retrieved 2016-12-02.
33. "Role of Nitrogen in Potato Production | N Deficiencies & Application | Yara". www.yara.us. Retrieved 2016-12-02.
34. Network, University of Nebraska-Lincoln | Web Developer. "Nitrogen | CropWatch". cropwatch.unl.edu. Retrieved 2016-12-02.
35. http://www.rainbowplantfood.com/agronomics/efu/nitrogen.pdf
36. http://www.upasitearesearch.org/soil-nutrition/
37. http://www.scribd.com/doc/15085476/CROP-NUTRIENT-DEFICIENCIES-TOXICITIES
38. http://digitalcommons.unl.edu/cgi/viewcontent.cgi?article=2349&context=extensionhist
39. Heinrich W. Scherer "Fertilizers" in Ullmann's Encyclopedia of Industrial Chemistry, 2000, Wiley-VCH, Weinheim. doi:10.1002/14356007.a10_323.pub3
40. International Plant Nutrition Institute. (1999). Functions of phosphorus in plants. Better crops with plant food, 83(1), 6-7.
41. Zambrosi, F. C. B., Ribeiro, R. V., Marchiori, P. E. R., Cantarella, H., & Landell, M. G. A. (2014). Sugarcane performance under phosphorus deficiency: physiological responses and genotypic variation. Plant and Soil, 386(1), 273–283.
42. Sawyer, J.E. (2008). Differentiating and Understanding the Mehlich 3, Bray, and Olsen Soil Phosphorus Tests. Retrieved from: http://www.agronext.iastate.edu/soilfertility/presentations/mbotest.pdf
43. Department of Soil Science, University of Wisconsin-Madison. (2004). Available phosphorus. Wisconsin Procedures for Soil Testing, Plant Analysis and Feed and Forage Analysis. Retrieved from: http://datcp.wi.gov/uploads/Farms/pdf/WIProcSoilTestingAnaysis.pdf.
44. Kowalenko, C.G & Babuin, D. (2007). Interference problems with phosphoantimonymolybdendum colorimetric measurement of phosphorus in

soil and plant materials. Communications in soil science and plant analysis, 38(9-10), 1299-1316.

45. Angelova, V., Bekjarov, G., Dospatliev, L., Ivanov, & K., Zaprjanova, P. (2010). ICP determination of phosphorus in soils and plants. Retrieved from: http://iuss.org/19th%20WCSS/Symposium/pdf/1629.pdf

46. Osborne, S. L., Schepers, J. S., Francis, D. D., & Schlemmer, M. R. (2002). Detection of Phosphorus and Nitrogen Deficiencies in Corn Using Spectral Radiance Measurements. Agronomy Journal,94(6), 1215–1221.

47. Beegle , D. & Durst, P.T. (2002). Managing phosphorus for crop production. Retrieved from: http://extension.psu.edu/plants/nutrient-management/educational/soil-fertility/managing-phosphorus-for-crop-production/extension_publication_file

48. Maathuis, F. J. (2009). Physiological functions of mineral macronutrients. Current Opinion in Plant Biology, 250–258.

49. Datnoff, L.E. et al. Mineral Nutrition and Plant Disease. The American Phytopathological Society, 2007 ISBN 0-89054-346-1

50. Leigh, R. A.; Wyn Jones, R. G. (1984). "A Hypothesis Relating Critical Potassium Concentrations for Growth to the Distribution and Functions of This Ion in the Plant Cell". New Phytologist. 97: 1–13. doi:10.1111/j.1469-8137.1984.tb04103.x. JSTOR 2434189.

51. Hopkins, W.G. and Huner, N.P.A. Introduction to Plant Physiology 4th edition

52. "Potassium deficiency in plants", 13 December 2001, 17 November 2010.

53. DEFRA (2010). Fertiliser manual (RB209) (8th ed.). Norwich: TSO. ISBN 9780112432869.

54. Alloway, B.J. (2008). Zinc in soils and crop nutrition (PDF). Brussels: International Zinc Association and International Fertilizer Industry Association. ISBN 9789081333108. Retrieved 14 April 2015.

55. Weir, R.G,; Cresswell, G.C.; Loebel, M.R. (1995). Plant nutrient disorders 2: Tropical fruit and nut crops. Melbourne: Inkata Press. ISBN 0909605904.

56. Weir, R.G.; Cresswell, G.C. (1993). Plant nutrition disorders: 3. Vegetable crops. Melbourne: Inkata Press. ISBN 0909605912.

57. Mengel, K.; Kirkby, E.A. (2001). Principles of plant nutrition. Dordrecht: Kluwer Academic Publishers. ISBN 079237150X.

58. Brown, P.H., I. Cakmak and Q. Zhang (1993) Form and function of zinc in plants. Chap 7 in Robson, A.D. (ed.) Zinc in Soils and Plants, Kluwer Academic Publishers, Dordrecht. pp 90-106.

CHAPTER 5

Plant Defense Against Herbivory

Chapter Contents

Plants must invest energy and nutrients to grow stems, leaves, roots, and reproductive tissues. When herbivores eat these tissues, plants must cope with decreased abilities to convert nutrients and energy into offspring. Therefore, natural selection favors plant traits that limit the negative effects of herbivores.

INTRODUCTION

Plant defense against herbivory or host-plant resistance (HPR) describes a range of adaptations evolved by plants which improve their survival and reproduction by reducing the impact of herbivores. Plants can sense being touched, and they can use several strategies to defend against damage caused by herbivores. Many plants produce secondary metabolites, known as allelochemicals, that influence the behavior, growth, or survival of herbivores. These chemical defenses can act as repellents or toxins to herbivores, or reduce plant digestibility.

Other defensive strategies used by plants include escaping or avoiding herbivores in any time and/or any place, for example by growing in a location where plants are not easily found or accessed by herbivores, or by changing seasonal growth patterns. Another approach diverts herbivores toward eating non-essential parts, or enhances the ability of a plant to recover from the damage caused by herbivory. Some plants encourage the presence of natural enemies of herbivores, which in turn protect the plant. Each type of defense can be either constitutive (always present in the plant), or induced (produced in reaction to damage or stress caused by herbivores).

Historically, insects have been the most significant herbivores, and the evolution of land plants is closely associated with the evolution of insects. While most plant defenses are directed against insects, other defenses have evolved that are aimed at vertebrate herbivores, such as birds and mammals. The study of plant defenses against herbivory is important, not only from an evolutionary view point, but also in the direct impact that these defenses have on agriculture, including human and livestock food sources; as beneficial 'biological control agents' in biological pest control programs; as well as in the search for plants of medical importance.

Foxgloves produce several deadly chemicals, namely cardiac and steroidal glycosides. Ingestion can cause nausea, vomiting, hallucinations, convulsions, or death.

Evolution of Defensive Traits

The earliest land plants evolved from aquatic plants around 450 million years ago (Ma) in the Ordovician period. Many plants have adapted to iodine-deficient terrestrial environment by removing iodine from their metabolism, in fact iodine is essential only for animal cells.An important antiparasitic action is caused by the block of the transport of iodide of animal cells inhibiting sodium-iodide symporter (NIS). Many plant pesticides are glycosides (as the cardiac digitoxin) and cyanogenic glycosides which liberate cyanide, which, blocking cytochrome c oxidase and NIS, is poisonous only for a large part of parasites and herbivores and not for the plant cells in which it seems useful in seed dormancy phase. Iodide is not pesticide, but is oxidized, by vegetable peroxidase, to iodine, which is a strong oxidant, it is able to kill bacteria, fungi and protozoa.

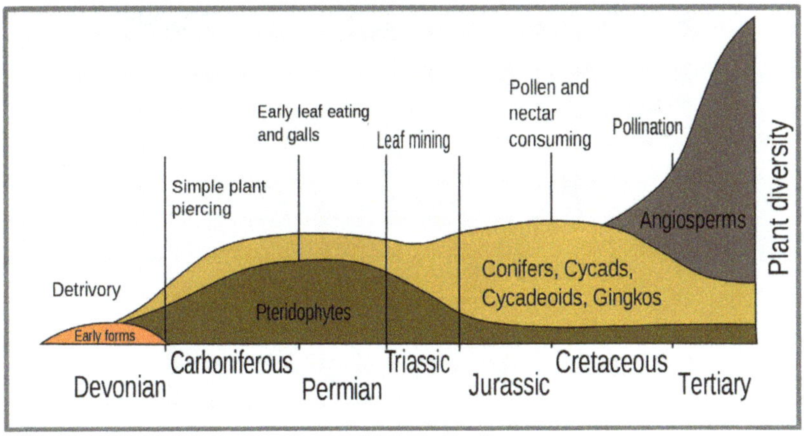

Timeline of plant evolution and the beginnings of different modes of insect herbivory

The Cretaceous period saw the appearance of more plant defense mechanisms. The diversification of flowering plants (angiosperms) at that time is associated with the sudden burst of speciation in insects. This diversification of insects represented a major selective force in plant evolution, and led to selection of plants that had defensive adaptations. Early insect herbivores were mandibulate and bit or chewed vegetation; but the evolution of vascular plants lead to the co-evolution of other forms of herbivory, such as sap-sucking, leaf mining, gall forming and nectar-feeding.

The relative abundance of different species of plants in ecological communities including forests and grasslands may be determined in part by the level of defensive compounds in the different species. Since the cost of replacement of damaged leaves is higher in conditions where resources are scarce, it may also be that plants growing in areas where water and nutrients are scarce may invest more resources into anti-herbivore defenses.

Records of Herbivores

Our understanding of herbivory in geological time comes from three sources: fossilized plants, which may preserve evidence of defense (such as spines), or herbivory-related damage; the observation of plant debris in fossilised animal faeces; and the construction of herbivore mouthparts.

Long thought to be a Mesozoic phenomenon, evidence for herbivory is found almost as soon as fossils which could show it. Within under 20 million years of the first fossils of sporangia and stems towards the close of the Silurian, around 420 million years ago, there is evidence that they were being consumed. Animals fed on the spores of early Devonian plants, and the Rhynie chert also provides evidence that organisms fed on plants using a "pierce and suck" technique. Many plants of this time are preserved with spine-like enations, which may have performed a defensive role before being co-opted to develop into leaves.

Viburnum lesquereuxii leaf with insect damage; Dakota Sandstone (Cretaceous) of Ellsworth County, Kansas. Scale bar is 10 mm.

During the ensuing 75 million years, plants evolved a range of more complex organs - from roots to seeds. There was a gap of 50 to 100 million years between each organ evolving, and it being fed upon. Hole feeding and skeletonization are recorded in the early Permian, with surface fluid feeding evolving by the end of that period.

A Plain Tiger *Danaus chrysippus* caterpillar making a moat to block defensive chemicals of *Calotropis* before feeding

Co-Evolution

Herbivores are dependent on plants for food, and have evolved mechanisms to obtain this food despite the evolution of a diverse arsenal of plant defenses. Herbivore adaptations to plant defense have been likened to *offensive traits* and consist of adaptations that allow increased feeding and use of a host plant. Relationships between herbivores and their host plants often results in reciprocal evolutionary change, called co-evolution. When an herbivore eats a plant it selects for plants that can mount a defensive response. In cases where this relationship demonstrates *specificity* (the evolution of each trait is due to the other), and *reciprocity* (both traits must evolve), the species are thought to have co-evolved. The "escape and radiation" mechanism for co-evolution presents the idea that adaptations in herbivores and their host plants have been the driving force behind speciation, and have played a role in the radiation of insect species during the age of angiosperms. Some herbivores have evolved ways to hijack plant defenses to their own benefit, by sequestering these chemicals and using them to protect themselves from predators. Plant defenses against herbivores are generally not complete so plants also tend to evolve some tolerance to herbivory.

HERBIVORE ADAPTATIONS TO PLANT DEFENSE

Herbivores are dependent on plants for food, and have coevolved mechanisms to obtain this food despite the evolution of a diverse arsenal of plant defenses against herbivory. Herbivore adaptations to plant defense have been likened to "offensive traits" and consist of those traits that allow for increased feeding and use of a host. Plants, on the other hand, protect their resources for use in growth and reproduction, by limiting the ability of herbivores to eat them. Relationships between herbivores and their host plants often results in reciprocal evolutionary change. When a herbivore eats a plant it selects for plants that can mount a defensive response, whether the response is incorporated biochemically or physically, or induced as a counterattack. In cases where this relationship demonstrates "specificity" (the evolution of each trait is due to the other), and "reciprocity" (both traits must evolve), the species are thought to have coevolved. The escape and radiation mechanisms for coevolution, presents the idea that adaptations in herbivores and their host plants, has been the driving force behind speciation. The coevolution that occurs between plants and herbivores that ultimately results in the speciation of both can be further explained by the Red Queen hypothesis. This hypothesis states that competitive success and failure evolve back and forth through organizational learning. The act of an organism facing competition with another organism ultimately leads to an increase in the organism's performance due to selection. This increase in competitive success then forces the competing organism to increase its performance through selection as well, thus creating an "arms race" between the two species. Herbivores evolve due to plant defenses because plants must increase their competitive performance first due to herbivore competitive success.

Mechanical Adaptations

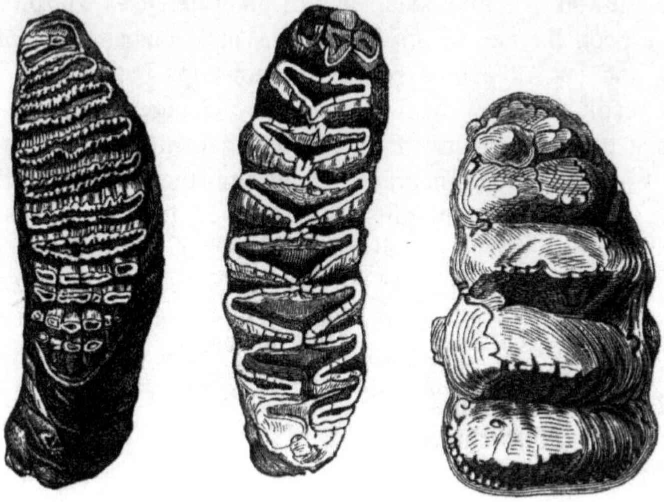

The molars of three species of elephant illustrate their different feeding preferences (l-asian elephant, c-african elephant, r-*Mastodon ginganteum*)

Herbivores have developed a diverse range of physical structures to facilitate the consumption of plant material. To break up intact plant tissues, mammals have developed teeth structures that reflect their feeding preferences. For instance, frugivores (animals that feed primarily on fruit) and herbivores that feed on soft foliage have low-crowned teeth specialized for grinding foliage and seeds. Grazing animals that tend to eat hard, silica-rich grasses, have high-crowned teeth, which are capable of grinding tough plant tissues and do not wear down as quickly as low-crowned teeth. Birds grind plant material or crush seeds using their beaks and gizzards.

Insect herbivores have evolved a wide range of tools to facilitate feeding. Often these tools reflect an individual's feeding strategy and its preferred food type.Within the family Sphingidae (sphinx moths), it has been observed that the caterpillars of species which eat relatively soft leaves are equipped with incisors for tearing and chewing, while the species that feed on mature leaves and grasses cut them with toothless snipping mandibles (the uppermost pair of jaws in insects, used for feeding).

A herbivore's diet often shapes its feeding adaptations. Grasshopper head size, and thus chewing power, was demonstrated to be greater for individuals raised on rye grass (a relatively hard grass) when compared to individuals raised on red clover (a soft diet). Larval lepidoptera that feed on plants with high levels of condensed tannins (as in trees) have more alkaline midguts when compared to lepidoptera that feed on herbs and forbs (pH of 8.67 vs. 8.29 respectively). This morphological difference can be explained by the fact that insoluble tannin-protein complexes can be broken down and absorbed as nutrients at alkaline pH levels.

Biochemical Adaptations

Herbivores generate enzymes that counter and reduce the effectiveness of numerous toxic secondary metabolic products produced by plants. One such enzyme group, mixed function oxidases (MFOs), detoxify harmful plant compounds by catalyzing oxidative reactions. Cytochrome P450 oxidases (or P-450), a specific class of MFO, have been specifically connected to detoxification of plant secondary metabolic products. One group linked herbivore feeding on plant material protected by chemical defenses with P-450 detoxification in larval tobacco hornworms. The induction of P-450 after initial nicotine ingestion allowed the larval tobacco hornworms to increase feeding on the toxic plant tissues.

An important enzyme produced by herbivorous insects is protease. The protease enzyme is a protein in the gut that helps the insect digest its main source of food: plant tissue. Many types of plants produce protease inhibitors, which inactivate proteases. Protease inactivation can lead to many issues such as reduced feeding, prolonged larval development time, and weight gain. However, many insects, including *S. exigua* and *L. decemlineatu* have been selected for mechanisms to avoid the effects of protease inhibitors. Some of these mechanisms include developing protease enzymes that are unaffected by the plant protease inhibitors, gaining the ability to degrade protease inhibitors, and acquiring mutations that allow the digesting of plant tissue without its destructive effects.

Herbivores may also produce salivary enzymes that reduce the degree of defense generated by a host plant. The enzyme glucose oxidase, a component of saliva for the caterpillar *Helicoverpa zea*, counteracts the production of induced defenses in tobacco. Similarly, aphid saliva reduces its host's induced response by forming a barrier between the aphid's stylet and the plant cells.

Behavioral Adaptations

Herbivores can avoid plant defenses by eating plants selectively in space and time. For the winter moth, feeding on oak leaves early in the season maximized the amount of protein and nutrients available to the moth, while minimizing the amount of tannins produced by the tree. Herbivores can also spatially avoid plant defenses. The piercing mouthparts of species in *Hemiptera* allow them to feed around areas of high toxin concentration. Several species of caterpillar feed on maple leaves by "window feeding" on pieces of leaf and avoiding the tough areas, or those with a high lignin concentration. Similarly, the cotton leaf perforator selectively avoids eating the epidermis and pigment glands of their hosts, which contain defensive terpenoid aldehydes. Some plants only produce toxins in small amounts, and rapidly deploy them to the area under attack. Some beetles counter this adaptation by attacking target plants in groups, thereby allowing each individual beetle to avoid ingesting too much toxin. Some animals ingest large amounts of poisons in their food, but then eat clay or other minerals, which neutralize the poisons. This behavior is known as geophagy.

Plant defense may explain, in part, why herbivores employ different life history strategies. Monophagous species (animals that eat plants from a single genus) must produce specialized enzymes to detoxify their food, or develop specialized structures to deal with sequestered chemicals. Polyphagous species (animals that eat plants from many different families), on the other hand, produce more detoxyfying enzymes (specifically MFO) to deal with a range of plant chemical defenses. Polyphagy often develops when a herbivore's host plants are rare as a necessity to gain enough food. Monophagy is favored when there is interspecific competition for food, where specialization often increases an animals' competitive ability to use a resource.

One major example of herbivorous behavioral adaptations deals with introduced insecticides and pesticides. The introduction of new herbicides and pesticides only selects for insects that can ultimately avoid or utilize these chemicals over time. Adding toxin free plants to a population of transgenic plants, or genetically modified plants that produce their own insecticides, has been shown to minimize the rate of evolution in insects feeding on crop plants. But even so, the rate of adaptation is only increasing in these insects.

Microbial Symbionts

Galls (upper left and right) A knopper gall formed on an acorn on the branch of an English Oak tree by the parthenogeneticgall wasp *Andricus quercuscalicis*.

Herbivores are unable to digest complex cellulose and rely on mutualistic, internal symbiotic bacteria, fungi, or protozoa to break down cellulose so it can be used by the herbivore. Microbial symbionts also allow herbivores to eat plants that would otherwise be inedible by detoxifying plant secondary metabolites. For example, fungal symbionts of cigarette beetles (*Lasioderma serricorne*) use certain plant allelochemicals as their source of carbon, in addition to producing detoxification enzymes (esterases) to get rid of other toxins. Microbial symbionts also assist in the acquisition of plant material by weakening a host plant's defenses. Some herbivores are more successful at feeding on damaged hosts. As an example, several species of bark beetle introduce blue stain fungi of the genera *Ceratocystis* and *Ophiostoma* into trees before feeding. The blue stain fungi cause lesions that reduce the trees' defensive mechanisms and allow the bark beetles to feed.

Host Manipulation

Herbivores often manipulate their host plants to use them better as resources. Herbivorous insects favorably alter the microhabitat in which the herbivore feeds to counter existing plant defenses. For example, caterpillars from the families Pyralidae and Ctenuchidae roll mature leaves of the neotropical shrub *Psychotria horizontalis* around an expanding bud that they consume. By rolling the leaves, the insects reduce the amount of light reaching the bud by 95%, and this shading prevents leaf toughness and leaf tannin concentrations in the expanding bud, while maintaining the amount of nutritional gain of nitrogen. Lepidoptera larvae also tie leaves together and feed on the inside of the leaves to decrease the effectiveness of the phototoxin hypericin in St. John's-wort.

Herbivores also manipulate their microhabitat by forming galls, plant structures made of plant tissue but controlled by the herbivore. Galls act as both domatia (housing), and food sources for the gall maker. The interior of a gall is composed of edible nutritious tissue. Aphid galls in narrow leaf cottonwood (*Populus angustifolia*) act as "physiologic sinks," concentrating resources in the gall from the surrounding plant parts. Galls may also provide the herbivore protection from predators.

Some herbivores use feeding behaviors that are capable of disarming the defenses of their host plants. One such plant defensive strategy is the use of latex and resin canals that contain sticky toxins and digestibility reducers. These canal systems store fluids under pressure, and when ruptured (i.e. from herbivory) secondary metabolic products flow to the release point.Herbivores can evade this defense, however, by damaging the leaf veins. This technique minimizes the outflow of latex or resin beyond the cut and allows herbivores to freely feed above the damaged section. Several strategies are employed by herbivores to relieve canal pressure, including vein cutting and trenching. The technique used by the herbivore corresponds to the architecture of the canal system. Dussourd and Denno examined the behavior of 33 species of insect herbivores on 10 families of plants with canals and found that herbivores on plants with branching canal systems used vein cutting, while herbivores found on plants with net-like canal systems employed trenching to evade plant defenses.

Herbivore Use of Plant Chemicals

Plant chemical defenses can be used by herbivores, by storing eaten plant chemicals, and using them in defense against predators. To be effective defensive agents, the sequestered chemicals cannot be metabolized into inactive products. Using plant chemicals can be costly to herbivores because it often requires specialized handling, storage, and modification. This cost can be seen when plants that use chemical defenses are compared to those plants that do not, in situations when herbivores are excluded. Several species of insects sequester and deploy plant chemicals for their own defense. Caterpillar and adult monarch butterflies store cardiac glycosides from milkweed, making these organisms distasteful. After eating a monarch caterpillar or butterfly, the bird predator will usually vomit, leading the bird to avoid eating similar looking butterflies in the future. Two different species of milkweed bug in the family Hemiptera, *Lygaeus kalmii* and Large Milkweed Bug *(Oncopeltus fasciatus)*, are colored with bright orange and black, and are said to be aposematically colored, in that they "advertise" their distastefulness by being brightly colored.

Monarch butterflies obtain poison from the plants they feed on as larvae, their distinctive appearance serving to warn predators.

Secondary metabolic products can also be useful to herbivores due to the antibiotic properties of the toxins, which can protect herbivores against pathogens. Additionally, secondary metabolic products can act as cues to identify a plant for feeding or oviposition (egg laying) by herbivores.

Types

Plant defenses can be classified generally as constitutive or induced. Constitutive defenses are always present in the plant, while induced defenses are produced or mobilized to the site where a plant is injured. There is wide variation in the composition and concentration of constitutive defenses and these range from mechanical defenses to digestibility reducers and toxins. Many external mechanical defenses and large quantitative defenses are constitutive, as they require large amounts of resources to produce and are difficult to mobilize. A variety of molecular and biochemical approaches are used to determine the mechanism of constitutive and induced plant defenses responses against herbivory.

Induced defenses include secondary metabolic products, as well as morphological and physiological changes. An advantage of inducible, as opposed to constitutive defenses, is that they are only produced when needed, and are therefore potentially less costly, especially when herbivory is variable.

TOXALBUMIN

Ricinus communis

Toxalbumins are toxic plant proteins that disable ribosomes and thereby inhibit protein synthesis, producing severe cytotoxic effects in multiple organ systems. They are dimers held together by a disulfide bond and comprise a lectin (carbohydrate-binding protein) part which binds to the cell membrane and enables the toxin part to gain access to the cell contents. Toxalbumins are similar in structure to the toxins found in cholera, tetanus, diphtheria and botulinum; and their physiological and toxic properties are similar to those of viperine snake venom.

Toxalbumins notably are present in the plant families Leguminosae and Euphorbiaceae, occurring for instance in *Robinia pseudoacacia*, *Abrus precatorius*, *Jatropha curcas*, *Croton gratissimus* and *Ricinus communis*. Typical toxalbumins are abrin and ricin. Ingestion of seed containing toxalbumins is not necessarily fatal as the hard seed coat will withstand digestion, unless the seed has been pierced, as would happen in the making of necklaces, prayer beads or bracelets, and even then the toxalbumin is likely to be digested and thereby rendered harmless. Toxalbumins injected intravenously or subcutaneously or inhaled in powdered form, though, are highly toxic. A latent period of hours to days may follow with no sensible signs of distress, after which symptoms of nausea, vomiting and diarrhoea will appear, followed by delirium, seizures, coma, and death. Seen from an evolutionary viewpoint, toxalbumins developed as a deterrent to consumption of seeds, foliage, bark and roots. Ripe fruits having a fleshy pulp are usually tasty and edible and lacking toxalbumins, encourage ingestion and the consequent distribution of seeds that have a coat sufficiently durable to survive their passage through the digestive system of a herbivore or fructivore.

Being soluble in water, ricin is not present in extracted oils. As with most proteins it breaks down after heat treatment, such as cooking or steaming, and after the oil is extracted, the resulting pomace is often used as animal feed. There is an enormous variation in sensitivity to the toxin, and a lethal dose may be as little as two-millionths of body weight. Since ricin is a protein, antibodies may be produced by inoculation, allowing resistance of up to 800 times a normal lethal dose. Ricin has been used in assassinations, a notorious case being the use of a 1.53 mm pellet holding a few hundred millionths of a gram of ricin to kill the Bulgarian broadcaster, Georgi Markov, who died 4 days after being attacked.

Toxalbumins were first described in about 1890 by Ludwig Brieger (1849–1919) and Sigmund Fraenkel (1868–1939), associates of the organic chemist Eugen Baumann. Brieger first used the term toxin.

The toxins present in poisonous mushrooms such as *Amanita phalloides* are quite different from toxalbumins and are mostly secondary metabolites or amatoxins which do not readily break down under applied heat. They are potent inhibitors of RNA polymerase II, an enzyme vital in the synthesis of messenger RNA (mRNA), microRNA, and small nuclear RNA (snRNA). Without mRNA, the template for protein synthesis, cell metabolism stops. In this respect their metabolic effect is similar to that of toxalbumins.

MUTUALISM (BIOLOGY)

Mutualism is the way two organisms of *different* species exist in a relationship in which each individual benefits from the activity of the other. Similar interactions *within* a species are known as co-operation. Mutualism can be contrasted with interspecific competition, in which each species experiences *reduced* fitness, and exploitation, or parasitism, in which one species benefits at the "expense" of the other. Symbiosis involves two species living in close proximity and includes relationships that are mutualistic, parasitic, and commensal. Symbiotic relationships are sometimes, but not always, mutualistic.

A well-known example of mutualism is the relationship between ungulates (such as bovines) and bacteria within their intestines. The ungulates benefit from the cellulase produced by the bacteria, which facilitates digestion; the bacteria benefit from having a stable supply of nutrients in the host environment. This can also be found in many many different symbiotic relationships.

Mutualism plays a key part in ecology. For example, mutualistic interactions are vital for terrestrial ecosystem function as more than 48% of land plants rely on mycorrhizal relationships with fungi to provide them with inorganic compounds and trace elements.

In addition, mutualism is thought to have driven the evolution of much of the biological diversity we see, such as flower forms (important for pollination mutualisms) and co-evolution between groups of species. However mutualism has historically received less attention than other interactions such as predation and parasitism.

Measuring the exact fitness benefit to the individuals in a mutualistic relationship is not always straightforward, particularly when the individuals can receive benefits from a variety of species, for example most plant-pollinator mutualisms. It is therefore common to categorise mutualisms according to the closeness of the association, using terms such as obligate and facultative. Defining "closeness," however, is also problematic. It can refer to mutual dependency (the species cannot live without one another) or the biological intimacy of the relationship in relation to physical closeness (*e.g.*, one species living within the tissues of the other species).

The term "mutualism" was introduced by Pierre-Joseph van Beneden in 1876.

Hummingbird hawkmoth drinking from Dianthus. Pollination is a classic example of mutualism.

Types of Relationships

Mutualistic transversals can be thought of as a form of "biological barter" in mycorrhizal associations between plant roots and fungi, with the plant providing carbohydrates to the fungus in return for primarily phosphate but also nitrogenous compounds. Other examples include rhizobia bacteria that fix nitrogen for leguminous plants (family Fabaceae) in return for energy-containing carbohydrates.

Service-Resource Relationships

Service-resource relationships are also common.

Pollination in which nectar or pollen (food resources) are traded for pollen dispersal (a service) or ant protection of aphids, where the aphids trade sugar-rich honeydew (a by-product of their mode of feeding on plant sap) in return for defense against predators such as ladybugs.

Phagophiles feed (resource) on ectoparasites, thereby providing anti-pest service, as in cleaning symbiosis. Elacatinus and Gobiosoma, genus of gobies, also feed on ectoparasites of their clients while cleaning them.

The red-billed oxpecker eats ticks on the impala's coat, in a cleaning symbiosis.

Zoochory is the dispersal of the seeds of plants by animals. This is similar to pollination in that the plant produces food resources (for example, fleshy fruit, overabundance of seeds) for animals that disperse the seeds (service).

Service-Service Relationships

Ocellaris clownfish and Ritter's sea anemones is a mutual service-service symbiosis, the fish driving off butterfly fish and the anemone's tentacles protecting the fish from predators.

Strict service-service interactions are very rare, for reasons that are far from clear. One example is the relationship between sea anemones and anemone fish in the family Pomacentridae: the anemones provide the fish with protection from predators (which cannot tolerate the stings of the anemone's tentacles) and the fish defend the anemones against butterflyfish (family Chaetodontidae), which eat anemones. However, in common with many mutualisms, there is more than one aspect to it: in the anemonefish-anemone mutualism, waste ammonia from the fish feed the symbioticalgae that are found in the anemone's tentacles. Therefore, what appears to be a service-service mutualism in fact has a service-resource component. A second example is that of the relationship between some ants in the genus *Pseudomyrmex* and trees in the genus *Acacia*, such as the whistling thorn and bullhorn acacia. The ants nest inside the plant's thorns. In exchange for shelter, the ants protect acacias from attack by herbivores (which they frequently eat, introducing a resource component to this service-service relationship) and competition from other plants by trimming back vegetation that would shade the acacia. In addition, another service-resource component is present, as the ants regularly feed on lipid-rich food-bodies called Beltian bodies that are on the *Acacia* plant.

In the neotropics, the ant, *Myrmelachista schumanni* makes its nest in special cavities in *Duroia hirsute*. Plants in the vicinity that belong to other species are killed with formic acid. This selective gardening can be so aggressive that small areas of the rainforest are dominated by *Duroia hirsute*. These peculiar patches are known by local people as "devil's gardens".

In some of these relationships, the cost of the ant's protection can be quite expensive. *Cordia* sp. trees in the Amazonian rainforest have a kind of partnership with *Allomerus* sp. ants, which make their nests in modified leaves. To increase the amount of living space available, the ants will destroy the tree's flower buds. The flowers die and leaves develop instead, providing the ants with more dwellings. Another type of *Allomerus* sp. ant lives with the *Hirtella* sp. tree in the same forests, but in this relationship the tree has turned the tables on the ants. When the tree is ready to produce flowers, the ant abodes on certain branches begin to wither and shrink, forcing the occupants to flee, leaving the tree's flowers to develop free from ant attack.

The term "species group" can be used to describe the manner in which individual organisms group together. In this non-taxonomic context one can refer to "same-species groups" and "mixed-species groups." While same-species groups are the norm, examples of mixed-species groups abound. For example, zebra (*Equus burchelli*) and wildebeest (*Connochaetes taurinus*) can remain in association during periods of long distance migration across the Serengeti as a strategy for thwarting predators. *Cercopithecus mitis* and *Cercopithecus ascanius*, species of monkey in the Kakamega Forest of Kenya, can stay in close proximity and travel along exactly the same routes through the forest for periods of up to 12 hours. These mixed-species groups cannot be explained by the coincidence of sharing the same habitat. Rather, they are created by the active behavioural choice of at least one of the species in question.

Humans

Humans also engage in mutualisms with other species, including their gut flora without which they would not be able to digest food efficiently. Infestations of head lice might have been beneficial for humans by fostering an immune response that helps to reduce the threat of body louse borne lethal diseases.

Some relationships between humans and domesticated animals and plants are to different degrees mutualistic. For example, agricultural varieties of maize provide food for humans and are unable to reproduce without human intervention because the leafy sheath does not fall open, and the seedhead (the "corn on the cob") does not shatter to scatter the seeds naturally.

In traditional agriculture, some plants have mutualist as companion plants, providing each other with shelter, soil fertility and/or natural pest control. For example, beans may grow up cornstalks as a trellis, while fixing nitrogen in the soil for the corn, a phenomenon that is used in Three Sisters farming.

Dogs and sheep were among the first animals to be domesticated.

Boran people of Ethiopia and Kenya traditionally use a whistle to call the honeyguide bird, though the practice is declining. If the bird is hungry and within earshot, it guides them to a bees' nest. In exchange the Borans leave some food from the nest for the bird.

A population of bottlenose dolphins in Laguna, Brazil coordinates, via body language, with local net-using fishermen in order for both to catch schools of mullet.

One researcher has proposed that the key advantage *Homo sapiens* had over Neanderthals in competing over similar habitats was the former's mutualism with dogs.

Mathematical Modeling

Mathematical treatments of mutualisms, like the study of mutualisms in general, has lagged behind those of predation, or predator-prey, consumer-resource, interactions. Here we present two such approaches. In models of mutualisms, the terms "type I" and "type II" functional responses refer to the linear and saturating relationships, respectively, between *benefit* provided to an individual of species 1 (y-axis) on the *density* of species 2 (x-axis).

Type I Functional Response

One of the simplest frameworks for modeling species interactions is the Lotka–Volterra equations. In this model, the change in population density of the two mutualists is quantified as:

$$\frac{dN}{dt} = r_1 N \left(1 - \frac{N}{K_1} + \beta_{12}\frac{M}{K_1}\right)$$

$$\frac{dM}{dt} = r_2 M \left(1 - \frac{M}{K_2} + \beta_{21}\frac{N}{K_2}\right)$$

where

- N and M = the population densities.
- r = intrinsic growth rate of the population.
- K = carrying capacity of its local environmental setting.
- β = coefficient converting encounters with one species to new units of the other.

Mutualism is in essence the logistic growth equation + mutualistic interaction. The mutualistic interaction term represents the increase in population growth of species one as a result of the presence of greater numbers of species two, and vice versa. As the mutualistic term is always positive, it may lead to unrealistic unbounded growth as it happens with the simple model. So, it is important to include a saturation mechanism to avoid the problem.

The type I functional response is visualized as the graph of $\frac{\beta_{12}}{K_1}M$ vs. M.

Type II Functional Response

In 1989, David Hamilton Wright modified the Lotka–Volterra equations by adding a new term, $\beta M/K$, to represent a mutualistic relationship. Wright also considered the concept of saturation, which means that with higher densities, there are decreasing benefits of further increases of the mutualist population. Without saturation, species' densities would increase indefinitely. Because that isn't possible due to environmental constraints and carrying capacity, a model that includes saturation would be more accurate. Wright's mathematical theory is based on the premise of a simple two-species mutualism model in which the benefits of mutualism become saturated due to limits posed by handling time. Wright defines handling time as the time needed to process a food item, from the initial interaction to the start of a search for new food items and assumes that processing of food and searching for food are mutually exclusive. Mutualists that display foraging behavior are exposed to the restrictions on handling time. Mutualism can be associated with symbiosis.

Handling time interactions In 1959, C. S. Holling performed his classic disc experiment that assumed the following: that (1), the number of food items captured is proportional to the allotted searching time; and (2), that there is a variable of handling time that exists separately from the notion of search time. He then developed an equation for the Type II functional response, which showed that the feeding rate is equivalent to

$$\frac{ax}{1+axT_H}$$

where,

- a = the instantaneous discovery rate
- x = food item density
- T_H = handling time

The equation that incorporates Type II functional response and mutualism is:

$$\frac{dN}{dt} = N\left[r(1-cN) + \frac{baM}{1+aT_H M}\right]$$

where

- *N* and *M* = densities of the two mutualists
- *r* = intrinsic rate of increase of *N*
- *c* = coefficient measuring negative intraspecific interaction. This is equivalent to inverse of the carrying capacity, 1/*K*, of *N*, in the logistic equation.
- *a* = instantaneous discovery rate
- *b* = coefficient converting encounters with *M* to new units of *N*

or, equivalently,

$$\frac{dN}{dt} = N[r(1-cN) + \beta M/(X+M)]$$

where

- $X = 1/a\ T_H$
- $\beta = b/T_H$

The model presented above is most effectively applied to free-living species that encounter a number of individuals of the mutualist part in the course of their existences. Of note, as Wright points out, is that models of biological mutualism tend to be similar qualitatively, in that the featured isoclines generally have a positive decreasing slope, and by and large similar isocline diagrams. Mutualistic interactions are best visualized as positively sloped isoclines, which can be explained by the fact that the saturation of benefits accorded to mutualism or restrictions posed by outside factors contribute to a decreasing slope.

The type II functional response is visualized as the graph of $\frac{baM}{1+aT_H M}$ vs. M.

Structure of Networks

Mutualistic networks made up out of the interaction between plants and pollinators were found to have a similar structure in very different ecosystems on different continents, consisting of entirely different species. The structure of these mutualistic networks may have large consequences for the way in which pollinator communities respond to increasingly harsh conditions and on the community carrying capacity.

Mathematical models that examine the consequences of this network structure for the stability of pollinator communities suggest that the specific way in which plant-pollinator networks are organized minimizes competition between pollinators,reduce the spread of indirect effects and thus enhance ecosystem stability and may even lead to strong indirect facilitation between pollinators when conditions are harsh. This means that pollinator species together can survive under harsh conditions. But it also means that pollinator species collapse simultaneously when conditions pass a critical point. This simultaneous collapse occurs, because pollinator species depend on each other when surviving under difficult conditions.

Such a community-wide collapse, involving many pollinator species, can occur suddenly when increasingly harsh conditions pass a critical point and recovery from such a collapse might not be easy. The improvement in conditions needed for pollinators to recover, could be substantially larger than the improvement needed to return to conditions at which the pollinator community collapsed.

PLANT USE OF ENDOPHYTIC FUNGI IN DEFENSE

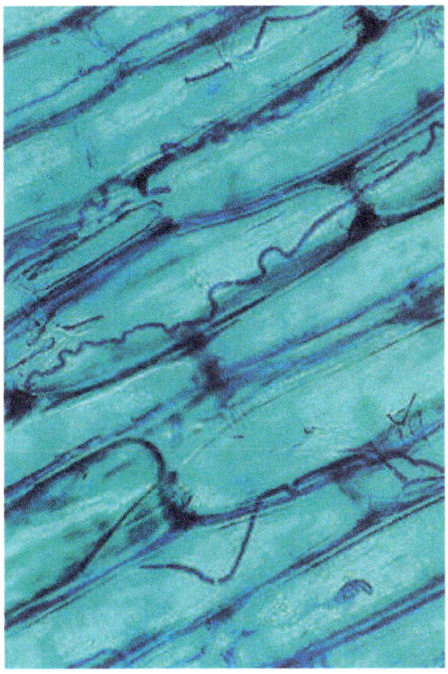

Neotyphodium spp.are commonly associated with tall fescue in the leaf sheath tissue. They produce secondary metabolites toxic to herbivores.

Plant use of endophytic fungi in defense occurs when endophytic fungi, which live symbiotically with the majority of plants by entering their cells, are utilized as an indirect defense against herbivores. In exchange for carbohydrate energy resources, the fungus provides benefits to the plant which can include increased water or nutrient uptake and protection from phytophagous insects, birds or mammals. Once associated, the fungi alter nutrient content of the plant and enhance or begin production of secondary metabolites. The change in chemical composition acts to deter herbivory by insects, grazing by ungulates and/or oviposition by adult insects. Endophyte-mediated defense can also be effective against pathogens and non-herbivory damage.

This differs from other forms of indirect defense in that the fungi live within the plant cells and directly alter their physiology. In contrast, other biotic defenses such as predators or parasites of the herbivores consuming a plant are normally attracted by volatile organic compounds (known as semiochemicals) released following damage or by food rewards and shelter produced by the plant. These defenders vary in the time spent with the plant: from long enough to oviposit to remaining there for numerous generations, as in the ant-acacia mutualism.Endophytic fungi tend to live with the plant over its entire life.

Diversity of Endophytic Associations

Claviceps spp. fungus growing on wheat spikes, a common endophyte of the grasses.

The fungal endophytes are a diverse group of organisms forming associations almost ubiquitously throughout the plant kingdom. The endophytes which provide indirect defense against herbivores may have come from a number of origins, including mutualistic root endophyte associations and the evolution of entomopathogenic fungi into plant-associated endophytes. The endomycorrhiza, which live in plant roots, are made up of five groups: arbuscular, arbutoid, ericoid, monotropoid, and orchid mycorrhizae. The majority of species are from the phylum Glomeromycota with the ericoid species coming from the Ascomycota, while the arbutoid, monotropoid and orchid mycorrhizae are classified as Basidiomycota. The entomopathogenic view has gained support from observations of increased fungal growth in response to induced plant defenses and colonization of plant tissues.

Examples of host specialists are numerous – especially in temperate environments – with multiple specialist fungi frequently infecting one plant individual simultaneously. These specialists demonstrate high levels of specificity for their host species and may form physiologically adapted host-races on closely related congeners. *Piriformospora indica* is an interesting endophytic fungus of the order Sebacinales, the fungus is capable of colonising roots and forming symbiotic relationship with every possible plant on earth . *P. indica* has also been shown to increase both crop yield and plant defence of a variety of crops(barley, tomato, maize etc.) against root-pathogens.However, there are also many examples of generalist fungi which may occur on different hosts at different frequencies (e.g. Acremonium endophytes from five subgenera of Festuca) and as part of a variety of fungal assemblages. They may even spread to novel, introduced plant species. Endophytic mutualists associate with species representative of every growth form and life history strategy in the grasses and many other groups of plants. The effects of associating with multiple strains or species of fungus at once can vary, but in general, one type of fungus will be providing the majority of benefit to the plant.

Mechanisms of Defense

Secondary Metabolite Production

Some chemical defenses once thought to be produced by the plant have since been shown to be synthesized by endophytic fungi. The chemical basis of insect resistance in endophyte-plant defense mutualisms has been most extensively studied in the perennial ryegrass and three major classes of secondary metabolites are found: indole diterpenes, ergot alkaloids and peramine. Related compounds are found across the range of endophytic fungal associations with plants. The terpenes and alkaloids are inducible defenses which act similarly to defensive compounds produced by plants and are highly toxic to a wide variety of phytophagous insects as well as mammalian herbivores. Peramine occurs widely in endophyte-associated grasses and may also act as a signal to invertebrate herbivores of the presence of more dangerous defensive chemicals. Terpenoids and ketones have been linked to protection from specialist and generalist herbivores (both insect and vertebrate) across the higher plants.

Generalist herbivores are more likely than specialists to be negatively affected by the defense chemicals that endophytes produce because they have, on average, less resistance to these specific, qualitative defenses. Among the chewing insects, infection by mycorrhizae can actually benefit specialist feeders even if it negatively affects generalists. The overall pattern of effects on insect herbivores seems to support this, with generalist mesophyll feeders experiencing negative effects of host infection, although phloem feeders appear to be affected little by fungal defenses.

Secondary metabolites may also affect the behaviour of natural enemies of herbivorous species in a multi-trophic defense/predation association. For instance, terpenoid production attracts natural enemies of herbivores to damaged plants. These enemies can reduce numbers of invertebrate herbivores substantially and may not be attracted in the absence of endophytic symbionts. Multi-trophic interactions can have cascading consequences for the entire plant community, with the potential to vary widely depending on the combination of fungal species infecting a given plant and the abiotic conditions.

Altered Nutrient Content

Due to the inherently nutrient-exchange based economy of the plant-endophyte association, it is not surprising that infection by fungi directly alters the chemical composition of plants, with corresponding impacts on their herbivores. Endophytes frequently increase apoplastic carbohydrate concentration, altering the C:N ratio of leaves and making them a less efficient source of protein. This effect can be compounded when the fungus also uses plant nitrogen to form N-based secondary metabolites such as alkaloids. For example, the thistle gall fly (*Urophora cardui*) experiences reduced performance on plants infected with endophytic fungi due to the decrease in N-content and ability to produce large quantities of high-quality gall tissue.Additionally, increased availability of limiting nutrients to plants improves overall performance and health, potentially increasing the ability of infected plants to defend themselves.

Impacts on Host Plants

Herbivory Prevention

Studies of fungal infection consistently reveal that plants with endophytes are less likely to suffer substantial damage, and herbivores feeding on infected plants are less productive. There are multiple modes through which endophytic fungi reduce insect herbivore damage, including avoidance (deterrence), reduced feeding, reduced development rate,reduced growth and/or population growth, reduced survival and reduced oviposition. Vertebrate herbivores such as birds, rabbits and deer show the same patterns of avoidance and reduced performance. Even below-ground herbivores such as nematodes and root-feeding insects are reduced by endophyte infection. The strongest evidence for anti-herbivore benefits of fungal endophytes come from studies of herbivore populations being extirpated when allowed to feed only on infected plants. Examples of local extinction have been documented in crickets, larval armyworms and flour beetles.

Yet chemical defenses produced by fungal endophytes are not universally effective, and numerous insect herbivores are unaffected by a given compound at one or more life history stages; larval stages are often more susceptible to toxins than adults. Even endophytes which purportedly provide some defense benefit to their hosts such as the Neotyphidium partner of many grass species in the alpine tundra do not always lead to avoidance or ill-effects on herbivores due to spatial variation in levels of consumption.

Mutualism-Pathogenicity Continuum

Not all endophytic symbioses confer protection from herbivores – only some species associations act as defense mutualisms. The difference between a mutualistic endophyte and a pathogenic one can be indistinct and dependent on interactions with other species or environmental conditions. Some fungi which are pathogens in the absence of herbivores may become beneficial under high levels of insect damage, such as species which kill plant cells in order to make nutrients available for their own growth, thereby altering nutritional content of leaves and making them a less desirable foodstuff.Some endomycorrhizae may provide defense benefits but at the cost of lost reproductive potential by rendering grasses partially sterile with their own fungal reproductive structures taking precedence. This is not unusual among fungi, as non-endophytic plant pathogens have similar conditionally beneficial effects on defense. Some species of endophyte may be beneficial for the plants in other ways (e.g. nutrient and water uptake) but will provide less benefit as a plant receives more damage and not produce defensive chemicals in response. The effect of one fungus on the plant can be altered when multiple strains of fungi are infecting a given individual in combination.

Some endomycorrhizae may actually promote herbivore damage by making plants more susceptible to it. For example, some oak fungal endophytes are positively correlated with the levels of damage from leaf miners (*Cameraria spp.*), although negatively correlated with number of larvae present due to a reduction of oviposition on infected plants, which partially mitigates the higher damage rate. This continuum between mutualism and pathogenicity of endophytic fungi has major implications for plant fitness depending on the species of partners available in a given environment; mutualist status is conditional in a way similar to pollination and can shift from one to the other just as frequently.

Fitness and Competitive Ability

Fungal endophytes which provide defensive services to their host plants may exert selective pressures favouring association through enhanced fitness relative to uninfected hosts. The fungus Neotyphodium spp. infects grasses and increases fitness under conditions with high levels of interspecific competition. It does this through a combination of benefits including anti-herbivore defenses and growth promoting factors. The customary assumption that plant growth promotion is the main way fungal mutualists improve fitness under attack from herbivores is changing; alteration of plant chemical composition and induced resistance are now recognized as factors of great importance in improving competitive ability and fecundity. Plants undefended by chemical or physical means at certain points in their life histories have higher survival rates when infected with beneficial endophytic fungi. The general trend of plants infected with mutualistic fungi outperforming uninfected plants under moderate to high herbivory exerts selection for higher levels of fungal association as herbivory levels increase.Unsurprisingly, low to moderate levels of herbivore damage also increases the levels of infection by beneficial endophytic fungi.

In some cases the symbiosis between fungus and plant reaches a point of inseparability; fungal material is transmitted vertically from the maternal parent plant to seeds, forming a near-obligate mutualism. Because seeds are an important aspect of both fecundity and competitive ability for plants, high germination rates and seedling survival increase lifetime fitness. When fitness of plant and fungus become tightly intertwined, it is in the best interest of the endophyte to act in a manner beneficial to the plant, pushing it further toward the mutualism end of the continuum. Such effects of seed defense can also occur in dense stands of conspecifics through horizontal transmission of beneficial fungi. Mechanisms of microbial association defense, protecting the seeds rather than the already established plants, can have such drastic impacts on seed survival that they have been recognized to be an important aspect of the larger 'seed defence theory'.

Climate Change

The range of associated plants and fungi may be altered as climate changes, and not necessarily in a synchronous fashion. Plants may lose or gain endophytes, with as yet unknown impacts on defense and fitness, although generalist species may provide indirect defense in new habitats more often than not. Above-ground and below-ground associations can be mutual drivers of diversity, so altering the interactions between plants and their fungi may also have drastic effects on the community at large, including herbivores. Changes in distribution may bring plants into competition with previously established local species, making the fungal community – and particularly the pathogenic role of fungus – important in determining outcomes of competition with non-native invasive species. As carbon dioxide levels rise, the amplified photosynthesis will increase the pool of carbohydrates available to endophytic partners, potentially altering the strength of associations. Infected C3 plants show greater relative growth rate under high CO2 conditions compared to uninfected plants, and it is possible that the fungi drive this pattern of increased carbohydrate production.

Levels of herbivory may also increase as temperature and carbon dioxide concentrations rise. However, should plants remain associated with their current symbiotic fungi, evidence suggests that the degree of defense afforded them should not be altered. Although the amount of damage caused by herbivores frequently increases under elevated levels of atmospheric CO2, the proportion of damage remains constant when host plants are infected by their fungal endophytes. The change in Carbon-Nitrogen ratio will also have important consequences for herbivores. As carbohydrate levels increase within plants, relative nitrogen content will fall, having the dual effects of reducing nutritional benefit per unit biomass and also lowering concentrations of nitrogen-based defenses such as alkaloids.

History of Research

Early Recognition

The effects of endophytic fungi on the chemical composition of plants have been known by humans for centuries in the form of poisoning and disease as well as medicinal uses. Especially noted were impacts on agricultural products and livestock.Recognition and study of the mutualism did not begin in earnest until the 1980s when early studies on the impacts of alkaloids on animal herbivory confirmed their importance as agents of deterrence. Biologists began to characterize the diversity of endophytic mutualists through primitive techniques such as isozyme analysis and measuring the effects of infection on herbivores. Basic descriptive accounts of these previously neglected species of fungus became a major goal for mycologists, and a lot of research focus shifted to associates of the grass family (Poaceae) in particular, because of the large number of species which represent economically important commodities to humans.

Recent Advances and Future Directions

In addition to continuing descriptive studies of the effects of infection by defense mutualist endophytes, there has been a sharp increase in the number of studies which delve further into the ecology of plant-fungus associations and especially their multi-trophic impacts. The processes by which endophytic fungi alter plant physiology and volatile chemical levels are virtually unknown, and limited current results show a lack of consistency under differing environmental conditions, especially differing levels of herbivory. Studies comparing the relative impacts of mutualistic endophytes on inducible defenses and tolerance show a central function of infection in determining both responses to herbivore damage. On the whole, molecular mechanisms behind endophyte-mediated plant defense has been an increasing focus of research over the past ten years.

Since the beginning of the biotechnology revolution, much research has been also focused on using genetically modified endophytes to improve plant yields and defensive properties. The genetic basis of response to herbivory is being explored in tall fescue, where it appears the production of jasmonic acid may play a role in downregulation of the host plant's chemical defense pathways when a fungal endophyte is present. In some cases, fungi that are closely associated with their hosts have transferred genes for secondary metabolite production to the host genome, which could help to explain multiple origins of chemical defenses within the phylogeny of various groups of plants. This represents an important line of inquiry to pursue, especially in regards to understanding the chemical pathways that can be utilized in biotechnological applications.

Importance to Humans

Agriculture and Livestock

The secondary chemicals produced by endophytic fungi when associated with their host plants can be very harmful to mammals including livestock and humans, causing more than 600 million dollars in losses due to dead livestock every year. For example, the ergot alkaloids produced by Claviceps spp. have been dangerous contaminants of rye crops for centuries. When not lethal, defense chemicals produced by fungal endophytes may lead to lower productivity in cows and other livestock feeding on infected forage. Reduced nutritional quality of infected plant tissue also lowers the performance of farm animals, compounding the effect of reduced feed uptake when provided with infected plant matter. Reduced frequency of pregnancy and birth has also been reported in cattle and horses fed with infected forage. Consequently, the dairy and meat-production industries must endure substantial economic losses.

Fungal resistance to herbivores represents an environmentally sustainable alternative to pesticides that has experienced reasonable success in agricultural applications. The organic farming industry has embraced mycorrhizal symbionts as one tool for improving yields and protecting plants from damage. Infected crops of soybean, ribwort plantain,cabbage, banana, coffee bean plant and tomato all show markedly lower rates of herbivore damage compared to uninfected plants. Endophytic fungi show great promise as a means of indirect biocontrol in large-scale agricultural applications. The potential for biotechnology to improve crop populations through inoculation with modified fungal strains could reduce toxicity to livestock and improve yields of human-consumed foods. The endophyte, either with detrimental genes removed or beneficial new genes added, is used as a surrogate host to transform the crops genetically. An endophyte of ryegrass has been genetically transformed in this way and used successfully to deter herbivores.

Understanding how to mediate top-down effects on crop populations caused by the enemies of herbivores as well as bottom-up effects of chemical composition in infected plants has important consequences for the management of agricultural industries. The selection of endophytes for agricultural use must be careful and consideration must be paid to the specific impacts of infection on all species of pest and predators or parasites, which may vary on a geographic scale. The union of ecological and molecular techniques to increase yield without sacrificing the health of the local or global environment is a growing area of research.

Pharmaceutical

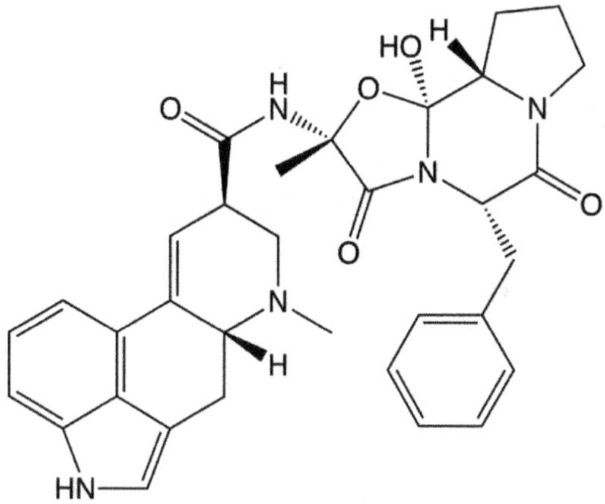

Ergotamine, a mycotoxin produced by Claviceps spp. which infects rye and related grasses, causing poisoning of livestock and humans. It is used in a number of medicinal

Many secondary metabolites from endophyte-plant interactions have also been isolated and used in raw or derived forms to produce a variety of drugs treating many conditions. The toxic properties of ergot alkaloids also make them useful in the treatment of headaches and throughout the process of giving birth by inducing contractions and stemming hemorrhages. Drugs used to treat Parkinson's Disease have been created from isolates of ergot toxins, although health risks may accompany their use.Ergotamine has also been used to synthesize lysergic acid diethylamide because of its chemical similarity to lysergic acid. The generally chemically-based defense properties of endophytic fungi make them a perfect group of organisms to search for new antibiotic compounds within, as other fungi have in the past yielded such useful drugs as penicillin and streptomycin and plants use their antibiotic qualities as a defense against pathogens.

THIGMONASTY

Thigmonasty or seismonasty is the nastic response of a plant or fungus to touch or vibration. Conspicuous examples of thigmonasty include many species in the leguminous subfamily Mimosoideae, active carnivorous plants such as Dionaea and a wide range of pollination mechanisms.

Mimosa pudica in normal and touched state.

Distinctive Aspects of Thigmonasty

Thigmonasty differs from thigmotropism in that nastic motion is independent of the direction of the stimulus. For example, tendrils from a climbing plant are thigmotropic because they twine around any support they touch, responding in whichever direction the stimulus came from. However, the shutting of a venus fly trap is thigmonastic; no matter what the direction of the stimulus, the trap simply shuts (and later possibly opens).

The time scales of thigmonastic responses tend to be shorter than those of thigmotropic movements because many examples of thigmonasty depend on pre-accumulated turgor or on bistable mechanisms rather than growth or cell division. Certain dramatic examples of rapid plant movement such as the sudden drooping of *Mimosa pudica* or the trapping action of *Dionaea* or *Utricularia* are fast enough to observe without time lapse photography; some take less than a second. Speed is no clear distinction however; for example the re-erection of *Mimosa* leaves is nastic, but typically takes some 15 to 30 minutes, rather than a second or so. Similarly, re-opening of the *Dionaea* trap, though also nastic, typically takes days to complete.

Botanical physiologists have discovered signalling molecules called turgorins, that help mediate the loss of turgor. In species with the fastest response time, vacuoles are believed to provide temporary, high speed storage for calcium ions.

Examples of Plants Exhibiting Thigmonasty

Thigmonasty in the Asteraceae

Thigmonasty other than leaf closure occurs in various species of thistles. When an insect lands on a flower, the anthers shrink and rebound, loading the insect with pollen. The effect results from turgor changes in specialized, highly elastic cell walls of the anthers. Similar pollination strategy occurs in *Rudbeckia hirta*.

Thigmonasty in the Droseraceae

The Venus Flytrap (*Dionaea muscipula*) presents a spectacular example of thigmonasty; when an insect lands on a trap formed by two curved lobes of a single leaf, the trap rapidly switches from an open to a closed configuration. Investigators have observed an action potential and changes in leaf turgor that accompany the reflex; they trigger the rapid elongation of individual cells. The common term for the elongation is acid growth although the process does not involve cell division.

Thigmonasty in the Fabaceae

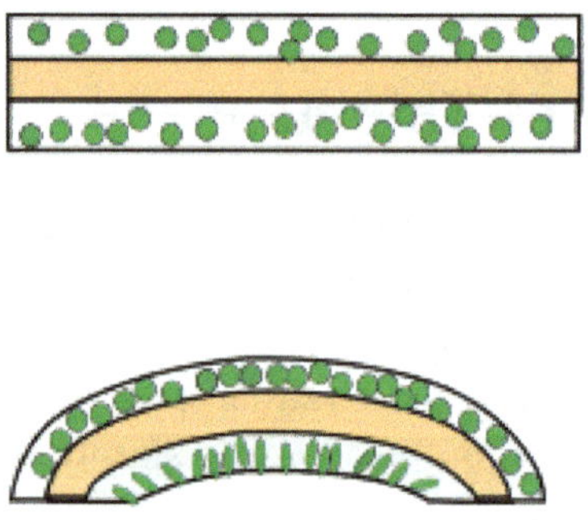

Pulvinus in extended and contracted position

Mimosa pudica is a plant with compound leaves that droop abruptly when stimulated. This is a classic example of thigmonastic action and has attracted detailed investigation. Contact or injury that causes leaflets to deform, will trigger an action potential. The action potential travels through the plant, initiating drooping of the leaflets as it passes. However, it cannot pass the pulvinus at the base of a petiole, and this prevents a local disturbance from causing all the leaves on the plant to collapse in response to trivial disturbances.

The pulvinus is a motor structure consisting of a rod of sclerenchyma surrounded by collenchyma. Such pulvini occur widely in the Fabaceae. In its extended position, the cells of the entire collar of collenchyma are distended with water. On receiving the action potential signal, the cells in the lower half of the pulvinus respond by expelling potassium and chlorine ions and taking up of calcium ions. This results in an osmotic gradient that draws water out of the affected cells, so that they temporarily shrink. This pulls the entire structure downward like a folding fan.

Many other Fabaceae react to touch with the same rapid leaf closure motion. These include the telegraph plant and the silk tree. The pea vine thigmonastically closes its leaves around a support. Catclaw Brier, a prairie mimosa, native to North America, shuts its leaves on contact. The plant is attractive to herbivores, and this behavior presumably provides protection against grazing.

Thigmonasty in the Oxalidaceae

Sensitive leaves also occur in plants of the wood sorrel family. Examples include many species of *Oxalis*, *Biophytum sensitivum*, and Averrhoa carambola (the plant which produces starfruit).

Other Forms

Some fungi exhibit trap closure similar to the venus fly trap. Mycologists have discovered action potentials in fungi but it is not currently clear whether they have any significance to thigmonastic behavior.

Optimal Defense Hypothesis

The optimal defense hypothesis attempts to explain how the kinds of defenses a particular plant might use reflect the threats each individual plant faces. This model considers three main factors, namely: risk of attack, value of the plant part, and the cost of defense.

The first factor determining optimal defense is risk: how likely is it that a plant or certain plant parts will be attacked? This is also related to the *plant apparency hypothesis*, which states that a plant will invest heavily in broadly effective defenses when the plant is easily found by herbivores. Examples of apparent plants that produce generalized protections include long-living trees, shrubs, and perennial grasses. Unapparent plants, such as short-lived plants of early successional stages, on the other hand, preferentially invest in small amounts of qualitative toxins that are effective against all but the most specialized herbivores.

The second factor is the value of protection: would the plant be less able to survive and reproduce after removal of part of its structure by a herbivore? Not all plant parts are of equal evolutionary value, thus valuable parts contain more defenses. A plant's stage of development at the time of feeding also affects the resulting change in fitness. Experimentally, the fitness value of a plant structure is determined by removing that part of the plant and observing the effect. In general, reproductive parts are not as easily replaced as vegetative parts, terminal leaves have greater value than basal leaves, and the loss of plant parts mid-season has a greater negative effect on fitness than removal at the beginning or end of the season. Seeds in particular tend to be very well protected. For example, the seeds of many edible fruits and nuts contain cyanogenic glycosides such as amygdalin. This results from the need to balance the effort needed to make the fruit attractive to animal dispersers while ensuring that the seeds are not destroyed by the animal. The final consideration is cost: how much will a particular defensive strategy cost a plant in energy and materials? This is particularly important, as energy spent on defense cannot be used for other functions, such as reproduction and growth. The optimal defense hypothesis predicts that plants will allocate more energy towards defense when the benefits of protection outweigh the costs, specifically in situations where there is high herbivore pressure.

Carbon:Nutrient Balance Hypothesis

The carbon:nutrient balance hypothesis, also known as the *environmental constraint hypothesis* or *Carbon Nutrient Balance Model* (CNBM), states that the various types of plant defenses are responses to variations in the levels of nutrients in the environment. This hypothesis predicts the Carbon/Nitrogen ratio in plants determines which secondary metabolites will be synthesized. For example, plants growing in nitrogen-poor soils will use carbon-based defenses (mostly digestibility reducers), while those growing in low-carbon environments (such as shady conditions) are more likely to produce nitrogen-based toxins. The hypothesis further predicts that plants can change their defenses in response to changes in nutrients. For example, if plants are grown in low-nitrogen conditions, then these plants will implement a defensive strategy composed of constitutive carbon-based defenses. If nutrient levels subsequently increase, by for example the addition of fertilizers, these carbon-based defenses will decrease.

Growth Rate Hypothesis

The growth rate hypothesis, also known as the *resource availability hypothesis*, states that defense strategies are determined by the inherent growth rate of the plant, which is in turn determined by the resources available to the plant. A major assumption is that available resources are the limiting factor in determining the maximum growth rate of a plant species. This model predicts that the level of defense investment will increase as the potential of growth decreases. Additionally, plants in resource-poor areas, with inherently slow-growth rates, tend to have long-lived leaves and twigs, and the loss of plant appendages may result in a loss of scarce and valuable nutrients.

A recent test of this model involved a reciprocal transplants of seedlings of 20 species of trees between clay soils (nutrient rich) and white sand (nutrient poor) to determine whether trade-offs between growth rate and defenses restrict species to one habitat. When planted in white sand and protected from herbivores, seedlings originating from clay outgrew those originating from the nutrient-poor sand, but in the presence of herbivores the seedlings originating from white sand performed better, likely due to their higher levels of constitutive carbon-based defenses. These finding suggest that defensive strategies limit the habitats of some plants.

Growth-Differentiation Balance Hypothesis

The growth-differentiation balance hypothesis states that plant defenses are a result of a tradeoff between "growth-related processes" and "differentiation-related processes" in different environments. Differentiation-related processes are defined as "processes that enhance the structure or function of existing cells (i.e. maturation and specialization)." A plant will produce chemical defenses only when energy is available from photosynthesis, and plants with the highest concentrations of secondary metabolites are the ones with an intermediate level of available resources. The GDBH also accounts for tradeoffs between growth and defense over a resource availability gradient. In situations where resources (e.g. water and nutrients) limit photosynthesis, carbon supply is predicted to limit both growth and defense. As resource availability increases, the requirements needed to support photosynthesis are met, allowing for accumulation of carbohydrate in tissues. As resources are not sufficient to meet the large demands of growth, these carbon compounds can instead be partitioned into the synthesis of carbon based secondary metabolites (phenolics, tannins, etc.). In environments where the resource demands for growth are met, carbon is allocated to rapidly dividing meristems (high sink strength) at the expense of secondary metabolism. Thus rapidly growing plants are predicted to contain lower levels of secondary metabolites and vice versa. In addition, the tradeoff predicted by the GDBH may change over time, as evidenced by a recent study on Salix spp. Much support for this hypothesis is present in the literature, and some scientists consider the GDBH the most mature of the plant defense hypotheses.

REFERENCES

1. Boyd, Jade (2012). "A bit touchy: Plants' insect defenses activated by touch". Rice University. http://news.rice.edu/2012/04/09/a-bit-touchy-plants-insect-defenses-activated-by-touch-2/
2. Venturi, S.; Donati, F.M.; Venturi, A.; Venturi, M. (2000). "Environmental Iodine Deficiency: A Challenge to the Evolution of Terrestrial Life?". Thyroid. 10 (8): 727–9. doi:10.1089/10507250050137851. PMID 11014322.
3. Venturi, Sebastiano (2011). "Evolutionary Significance of Iodine". Current Chemical Biology-. 5 (3): 155–162. doi:10.2174/187231311796765012. ISSN 1872-3136.
4. Ehrlich, Paul R.; Peter H. Raven (December 1964). "Butterflies and plants: a study of coevolution.". Evolution. Society for the Study of Evolution. 18 (4): 586–608. doi:10.2307/2406212. JSTOR 2406212.
5. Labandeira, C.C.; D.L. Dilcher, D.R. Davis, D.L. Wagner; Davis, D. R.; Wagner, D. L. (1994). "Ninety-seven million years of angiosperm-insect association: paleobiological insights into the meaning of coevolution" (PDF). Proceedings of the National Academy of Sciences of the United States of America. 91 (25): 12278–82. Bibcode:1994PNAS...9112278L. doi:10.1073/pnas.91.25.12278. PMC 45420 . PMID 11607501.
6. Keddy, P.A. 2007. Plants and Vegetation: Origins, Processes, Consequences. Cambridge University Press, Cambridge, UK. 666 p. Chapter 7.

7. Labandeira, C.C. (1998). "Early History Of Arthropod And Vascular Plant Associations 1". Annual Review of Earth and Planetary Sciences. 26 (1): 329–377. Bibcode:1998AREPS..26..329L. doi:10.1146/annurev.earth.26.1.329.
8. Labandeira, C. (2007). "The origin of herbivory on land: Initial patterns of plant tissue consumption by arthropods". Insect Science. 14 (4): 259–275. doi:10.1111/j.1744-7917.2007.00152.x.
9. Karban, Richard; Anurag A. Agrawal (November 2002). "Herbivore offense". Annual Review of Ecology and Systematics. 33 (1): 641–664. doi:10.1146/annurev.ecolsys.33.010802.150443.
10. Futuyma, Douglas J.; Montgomery Slatkin (1983). Coevolution. Sunderland, Massachusetts: Sinauer Associates. ISBN 0-87893-228-3.
11. Thompson, J. (1999). "What we know and do not know about coevolution: insect herbivores and plants as a test case.". In H. Olff; V. K. Brown; R. H. Drent. Herbivores: between plants and predators; the 38th symposium of the British Ecological Society in cooperation with the Netherlands Ecological Society held at the Wageningen Agricultural University, The Netherlands, 1997. Oxford: Blackwell Science. pp. 7–30. ISBN 0-632-05155-8.
12. Farrell, Brian D.; Charles Mitter (1994). "Adaptive Radiation in Insects and Plants: Time and Opportunity". American Zoologist. 34 (1): 57–69. doi:10.1093/icb/34.1.57.
13. Traw, Brian M.; Todd E. Dawson (May 2002). "Differential induction of trichomes by three herbivores of black mustard" (PDF). Oecologia. 131 (4): 526–532. doi:10.1007/s00442-002-0924-6. Retrieved 2007-05-27.
14. Walling, L.L. (2000). "The myriad plant responses to herbivores". J. Plant Growth Regul. 19: 195–216.
15. Wu, J.; Baldwin, I.T. (2009). "Herbivory-induced signalling in plants: Perception and action". Plant Cell Environ. 32: 1161–1174. doi:10.1111/j.1365-3040.2009.01943.x.
16. Sarmento, R.A.; Lemos, F.; Dias, C.R.; Kikuchi, W.T.; Rodrigues, J.C.P.; Pallini, A.; Sabelis, M.W.; Janssen, A. "A herbivorous mite down-regulates plant defence and produces web to exclude competitors. PLoS One 2011; 6, doi:10.1371/journal.pone.0023757
17. Sangha, J.S.; Yolanda; Chen, H.; Kaur, Jatinder; Khan, Wajahatullah; Abduljaleel, Zainularifeen; Alanazi, Mohammed S.; Mills, Aaron; Adalla, Candida B.; Bennett, John; Prithiviraj, Balakrishnan; Jahn, Gary C.; Leung, Hei (2013). "Proteome Analysis of Rice (Oryza sativa L.) Mutants Reveals Differentially Induced Proteins during Brown Planthopper (Nilaparvata lugens) Infestation". Int. J. Mol. Sci. 14: 3921–3945. doi:10.3390/ijms14023921. PMC 3588078 . PMID 23434671.
18. Karban, Richard; Anurag A. Agrawal; Marc Mangel (July 1997). "The benefits of induced defenses against herbivores" (PDF). Ecology. Ecological Society of America. 78 (5): 1351–1355. doi:10.2307/2266130. JSTOR 2266130. Retrieved 2007-05-27.
19. Fraenkel, G. (1959). "The raison d'être of secondary plant substances".

Science. 129 (3361): 1466–70. Bibcode:1959Sci...129.1466F. doi:10.1126/science.129.3361.1466. PMID 13658975.

20. Whittaker, Robert H. (1970). "The biochemical ecology of higher plants". In Ernest Sondheimer; John B. Simeone. Chemical ecology. Boston: Academic Press. pp. 43–70. ISBN 0-12-654750-5.
21. Whittaker, Robert H. (1975). Communities and ecosystems. New York: Macmillan. ISBN 0-02-427390-2.
22. Carmona, Diego; Marc J. Lajeunesse; Marc T.J. Johnson (April 2011). "Plant traits that predict resistance to herbivores" (PDF). Functional Ecology. 25 (2): 358–367. doi:10.1111/j.1365-2435.2010.01794.x. Retrieved 26 June 2011.
23. Theis, Nina; Manuel Lerdau (2003). "The evolution of function in plant secondary metabolites" (PDF). International Journal of Plant Science. 164 (3 Suppl.): S93–S102. doi:10.1086/374190. Archived from the original (PDF) on 2007-04-18. Retrieved 2007-05-27.
24. "Geraniums and Begonias". Retrieved 2014-01-15.
25. "Biochemical defenses: secondary metabolites". Plant Defense Systems & Medicinal Botany. Retrieved 2007-05-21.
26. "Alkaloids: contain a N-containing heterocycle". Plant Defense Systems & Medicinal Botany. Retrieved 2007-06-26.
27. Roberts, Margaret F.; Michael Wink (1998). Alkaloids: biochemistry, ecology, and medicinal applications. New York: Plenum Press. ISBN 0-306-45465-3.
28. Sneden, Albert T. "Alkaloids". Natural Products as Medicinally Useful Agents. Retrieved 2007-05-21.
29. Rhoades, David F (1979). "Evolution of Plant Chemical Defense against Herbivores". In Rosenthal, Gerald A.; Janzen, Daniel H. Herbivores: Their Interaction with Secondary Plant Metabolites. New York: Academic Press. pp. 3–54. ISBN 0-12-597180-X.
30. Toxicon Volume 38, Issue 1, January 2000, Pages 11-36 János Vetter Plant cyanogenic glycosides doi:10.1016/S0041-0101(99)00128-2
31. Niemeyer, HM (2009). "Hydroxamic acids derived from 2-hydroxy-2H-1,4-benzoxazin-3(4H)-one: key defense chemicals of cereals". J Agric Food Chem. 57 (5): 1677–1696. doi:10.1021/jf8034034.
32. Glauser, G; Marti, G; Villard, N; Doyen, GA; Wolfender, J-L; Turlings, TCJ; Erb, M (2011). "Induction and detoxification of maize 1,4-benzoxazin-3-ones by insect herbivores". Plant Journal. 68 (5): 901–911. doi:10.1111/j.1365-313X.2011.04740.x.
33. "Terpenoids". Plant Defense Systems & Medicinal Botany. Retrieved 2007-06-26.
34. Gershenzon, Jonathan; Wolfgang Kreis (1999). "Biochemistry of terpenoids". In Michael Wink. Biochemistry of plant secondary metabolism. London: Sheffield Academic Press. pp. 222–279. ISBN 0-8493-4085-3.

35. Sneden, Albert T. "Terpenes". Natural Products as Medicinally Useful Agents. Retrieved 2007-05-21.
36. "Phenols". Plant Defense Systems & Medicinal Botany. Retrieved 2007-05-21.
37. Van Soest, Peter J. (1982). Nutritional ecology of the ruminant: ruminant metabolism, nutritional strategies, the cellulolytic fermentation, and the chemistry of forages and plant fibers. Corvallis, Oregon: O & B Books. ISBN 0-9601586-0-X.
38. John W. Hylin (1969). "Toxic peptides and amino acids in foods and feeds". Journal of Agricultural and Food Chemistry. 17 (3): 492–496. doi:10.1021/jf60163a003.
39. E. Anet; B. Lythgoe; M. H. Silk; S. Trippett (1953). "Oenanthotoxin and cicutoxin. Isolation and structures". Journal of the Chemical Society: 309–322. doi:10.1039/JR9530000309.
40. Mark V. Barrow; Charles F. Simpson; Edward J. Miller (1974). "Lathyrism: A Review". The Quarterly Review of Biology. 49 (2): 101–128. doi:10.1086/408017. JSTOR 2820941. PMID 4601279.
41. Donald A. Levin; King, Dennis R. (1991). "The Impact of Fluoroacetate-Bearing Vegetation on Native Australian Fauna: A Review". Oikos. Blackwell Publishing. 61 (3): 412–430. doi:10.2307/3545249. JSTOR 3545249.
42. Steinberg, Peter D. (1986). "Chemical defenses and the susceptibility of tropical marine brown algae to herbivores". Oecologia. 69 (4): 628–630. doi:10.1007/BF00410374.
43. Fernandes GW (1994). "Plant mechanical defenses against insect herbivory". Revista Brasileira de Entomologia. 38 (2): 421–433 .
44. Cooper, Susan M.; Norman Owen-Smith (September 1986). "Effects of plant spinescence on large mammalian herbivores". Oecologia. 68 (3): 446–455. doi:10.1007/BF01036753.
45. Bond W, Lee W & Craine J (2004). "Plant structural defences against browsing birds: a legacy of New Zealand's extinct moas". Oikos. 104 (3): 500–508. doi:10.1111/j.0030-1299.2004.12720.x.
46. Young, Truman P. (1987). "Increased thorn length in Acacia drepanolobium-an induced response to browsing". Oecologia. 71 (3): 436–438. doi:10.1007/BF00378718.
47. Young, Truman P.; Bell Okello (1998). "Relaxation of an induced defense after exclusion of herbivores: spines on Acacia drepanolobium". Oecologia. 115 (4): 508–513. doi:10.1007/s004420050548.
48. Raven, Peter H.; Ray F. Evert; Susan E. Eichhorn (2005). Biology of Plants. New York: W. H. Freeman and Company. ISBN 0-7167-1007-2.
49. Williams, Kathy S.; Lawrence E. Gilbert (April 1981). "Insects as selective agents on plant vegetative morphology: egg mimicry reduces egg-laying by butterflies". Science. 212 (4493): 467–469. Bibcode:1981Sci...212..467W. doi:10.1126/science.212.4493.467. PMID 17802547.

50. Dicke, Marcel; Joop J.A. van Loon (December 2000). "Multitrophic effects of herbivore-induced plant volatiles in an evolutionary context". Entomologia Experimentalis et Applicata. 97 (3): 237–249. doi:10.1046/j.1570-7458.2000.00736.x.

51. Allmann, S.; Baldwin, I. T. (2010). "Insects Betray Themselves in Nature to Predators by Rapid Isomerization of Green Leaf Volatiles". Science. 329 (5995): 1075–8. Bibcode:2010Sci...329.1075A. doi:10.1126/science.1191634. PMID 20798319.

52. Schuman, Meredith C.; Barthel, Kathleen; Baldwin, Ian T. (October 2012). "Herbivory-induced volatiles function as defenses increasing fitness of the native plant Nicotiana attenuata in nature" (PDF). eLife Sciences. 1. doi:10.7554/eLife.00007.

53. Heil, Martin; Brigitte Fiala, K. Eduard Linsenmair, Gerhard Zotz, Petra Menke (December 1997). "Food body production in Macaranga triloba (Euphorbiaceae): A plant investment in anti-herbivore defense via symbiotic ant partners". Journal of Ecology. British Ecological Society. 85 (6): 847–861. doi:10.2307/2960606. JSTOR 2960606.

54. Young, Truman P.; Cynthia H. Stubblefield; Lynne A. Isbell (January 1997). "Ants on swollen-thorn acacias: species coexistence in a simple system". Oecologia. 109 (1): 98–107. doi:10.1007/s004420050063.

55. Williams, Alan G.; Thomas G. Whitham (December 1986). "Premature Leaf Abscission: An Induced Plant Defense Against Gall Aphids". Ecology. Ecological Society of America. 67 (6): 1619–1627. doi:10.2307/1939093. JSTOR 1939093.

56. Lev-Yadun, Simcha; Amots Dafni; Moshe A. Flaishman; Moshe Inbar; Ido Izhaki; Gadi Katzir; Gidi Ne'eman (October 2004). "Plant coloration undermines herbivorous insect camouflage" (PDF). BioEssays. 26 (10): 1126–1130. doi:10.1002/bies.20112. PMID 15382135. Retrieved 2007-05-27.

57. Archetti, M., M (2000). "The origin of autumn colours by coevolution". J. Theor. Biol. 205 (4): 625–630. doi:10.1006/jtbi.2000.2089. PMID 10931756.

58. Hamilton, W. D.; Brown, S. P. (2001). "Autumn tree colours as a handicap signal" (PDF). Proc. R. Soc. B. 268 (1475): 1489–1493. doi:10.1098/rspb.2001.1672. PMC 1088768 . PMID 11454293.

59. Stamp, Nancy (March 2003). "Out of the quagmire of plant defense hypotheses". Quarterly Review of Biology. 78 (1): 23–55. doi:10.1086/367580. PMID 12661508.

60. Rhoades, D. F.; R. G. Cates. (1974). "Towards a general theory of plant antiherbivore chemistry". In V. C. Runeckles; E. E. Conn. Recent advances in phytochemistry: proceedings of the annual meeting of the Phytochemical society of North America. Boston: Academic Press. pp. 168–213. ISBN 0-12-612408-6.

61. Wilf, Peter; Conrad C. Labandeira; Kirk R. Johnson; Phyllis D. Coley; Asher D. Cutter (2001). "Insect herbivory, plant defense, and early Cenozoic climate change" (PDF). Proceedings of the National Academy of Sciences. 98 (11): 6221–6226. Bibcode:2001PNAS...98.6221W. doi:10.1073/pnas.111069498. PMC 33449 . PMID 11353840. Retrieved 2007-05-27.

62. Feeny, P. (1976). "Plant apparency and chemical defense.". In James W. Wallace; Richard L. Mansell. Biochemical interaction between plants and insects: proceedings of the fifteenth annual meeting of the Phytochemical Society of North America. New York: Plenum Press. pp. 1–40. ISBN 0-306-34710-5.
63. D., McKey (1979). "The distribution of secondary compounds within plants.". In Gerald A. Rosenthal; Daniel H. Janzen. Herbivores, their interaction with secondary plant metabolites. Boston: Academic Press. pp. 55–133. ISBN 0-12-597180-X.
64. Krischik, V. A.; R. F. Denno (1983). "Individual, population, and geographic patterns in plant defense.". In Robert F. Denno; Mark S. McClure. Variable plants and herbivores in natural and managed systems. Boston: Academic Press. pp. 463–512. ISBN 0-12-209160-4.
65. Zangerl, Arthur R.; Claire E. Rutledge (April 1996). "The probability of attack and patterns of constitutive and induced defense: A test of optimal defense theory". The American Naturalist. 147 (4): 599–608. doi:10.1086/285868. JSTOR 2463237. Retrieved 2014-01-15.
66. Swain, Elisabeth; Chun Ping Li; Jonathan E. Poulton (1992). "Development of the Potential for Cyanogenesis in Maturing Black Cherry (Prunus serotina Ehrh.) Fruits". Plant Physiology. 98 (4): 1423–1428. doi:10.1104/pp.98.4.1423. PMC 1080367 . PMID 16668810.
67. Witmer, M.C. (1998). "Ecological and evolutionary implications of energy and protein requirements of avian frugivores eating sugary diets". Physiological Zoology. 71 (6): 599–610. PMID 9798248.
68. Pennings, Steven C.; Erin L. Siska; Mark D. Bertness (May 2001). "Latitudinal differences in plant palatability in Atlantic coast salt marshes". Ecology. Ecological Society of America. 82 (5): 1344–1359. doi:10.2307/2679994. JSTOR 2679994.
69. Bryant, John P.; Stuart Chapin, III; David R. Klein (May 1983). "Carbon/nutrient balance of boreal plants in relation to vertebrate herbivory". Oikos. Blackwell Publishing. 40 (3): 357–368. doi:10.2307/3544308. JSTOR 3544308.
70. Tuomi, J.; P. Niemela; F. S. Chapin, III; J. P. Bryant; S. Siren. (1988). "Defensive responses of trees in relation to their carbon/nutrient balance.". In William J. Mattson; Jean Levieux; C. Bernard-Dagan. Mechanisms of woody plant defenses against insects: search for pattern. Berlin: Springer-Verlag. pp. 57–72. ISBN 0-387-96673-0.
71. Colley, Phyllis D.; John P. Bryant; F. Stuart Chapin III (1985). "Resource availability and plant antiherbivore defense". Science. 230 (4728): 895–899. Bibcode:1985Sci...230..895C. doi:10.1126/science.230.4728.895. PMID 17739203.
72. Chapin, F. Stuart, III (1980). "The Mineral Nutrition of Wild Plants". Annual Review of Ecology and Systematics. 11: 233–260. doi:10.1146/annurev.es.11.110180.001313. JSTOR 2096908. Retrieved 2014-01-15.
73. Fine, Paul V. A.; Italo Mesones; Phyllis D. Coley (July 2004). "Herbivores promote habitat specialization by trees in Amazonian forests". Science. 305 (5684): 663–5. Bibcode:2004Sci...305..663F. doi:10.1126/science.1098982. PMID

15286371.

74. Loomis, W. E. (1981). "Growth and differentiation—an introduction and summary.". In P. F. Wareing; I. D. J. Phillips. Growth and differentiation in plants. New York: Pergamon Press. pp. 1–17. ISBN 0-08-026351-8.
75. Herms, Daniel A.; William J. Mattson (September 1992). "The dilemma of plants: to grow or defend". Quarterly Review of Biology. 67 (3): 283–335. doi:10.1086/417659. JSTOR 2830650. Retrieved 2014-01-15.
76. Polavarapu, Sridhar (2001). "Plant Resistance to insects". Agricultural Entomology & Pest Management. Rutgers University. Retrieved 2007-05-16.
77. Painter, Reginald Henry (1951). Insect Resistance in Crop Plants. Lawrence: University of Kansas Press. OCLC 443998.
78. Chesnokov, Pavel G. (1953). Methods of Investigating Plant Resistance to Pests. Jerusalem: Israel Program for Scientific Translations. OCLC 3576157.
79. Gleadow, Roslyn M.; Ian E. Woodrow (2002). "Constraints on effectiveness of cyanogenic glycosides in herbivore defense". Journal of Chemical Ecology. 28 (7): 1301–13. doi:10.1023/A:1016298100201. PMID 12199497.
80. Vough, Lester R.; E. Kim Cassel (July 2002). "Prussic Acid Poisoning of Livestock: Causes and Prevention (ExEx 4016)" (PDF). Extension Extra. South Dakota State University Extension Service.
81. Grant, G; Linda J. More; Norma H. McKenzie; Arpad Pusztai (1982). "The effect of heating on the haemagglutinating activity and nutritional properties of bean (Phaseolus vulgaris) seeds". Journal of the Science of Food and Agriculture. 33 (12): 1324–6. doi:10.1002/jsfa.2740331220. PMID 7166934.
82. Tu Jean-Louis (1999). "Natural Toxins in Raw Foods and How Cooking Affects Them". Is Cooked Food Poison?. Beyond Vegetarianism. Retrieved 2007-05-22.
83. George W. (2004). The Pesticide Book. Willoughby: MeisterPro. ISBN 1-892829-11-8. firat Ware
84. Michael Smith, C. (2005). Plant Resistance to Arthropods: Molecular and Conventional Approaches. Berlin: Springer. ISBN 1-4020-3701-5.
85. Dethier, V. G. (March 1954). "Evolution of feeding preferences in phytophagous insects". Evolution. Society for the Study of Evolution. 8 (1): 33–54. doi:10.2307/2405664. JSTOR 2405664.
86. Russ, Karen. "Less toxic insecticides" (PDF). Clemson University Home & Garden Information Center. Retrieved 2007-05-27.
87. "Linden, J., Stoner, R., Knutson, K. Gardner-Hughes, C. "Organic Disease Control Elicitors". Agro Food Industry Hi-Te (p12-15 Oct 2000)" (PDF).
88. van Emden, H.F. (November 1999). "Transgenic Host Plant Resistance to Insects—Some Reservations". Annals of the Entomological Society of America. 92 (6): 788–797. Retrieved 2007-05-27.
89. Ghosh, B. (2000). "Polyamines and plant alkaloids". Indian Journal of Experimental Biology. 38 (11): 1086–91. PMID 11395950.

90. Eisner, Thomas (March 1990). "Prospecting for nature's chemical riches". Chemoecology. 1 (1): 38–40. doi:10.1007/BF01240585.
91. Albert T. Sneden. "The Quinine Alkaloids" (PDF). Medicinal Chemistry and Drug Design. Retrieved 2007-05-23.
92. Albert T. Sneden. "The Tropane Alkaloids" (PDF). Medicinal Chemistry and Drug Design. Retrieved 2007-05-23.
93. Albert T. Sneden. "Taxol (Paclitaxe)" (PDF). Medicinal Chemistry and Drug Design. Retrieved 2007-05-23.
94. Hunter, William (1891). "On the Nature, Action, and Therapeutic Value of the Active Principles of Tuberculin". British Medical Journal. 2 (1595): 169–76. doi:10.1136/bmj.2.1595.169. JSTOR 20243302. PMC 2273336 . PMID 20753375.
95. "Abrus Precatorius - Medicinal Uses, Benefits of Abrus". Planetayurveda.com. Retrieved 2015-06-16.
96. "Toxalbumins - Peas and beans gone bad". Drugs and Poisons. 2004-02-26. Retrieved 2015-06-16.[self-published source?]
97. Brent, Jeffrey (2005). Critical Care Toxicology: Diagnosis and Management of the Critically Poisoned Patient. Mosby. p. 1345. ISBN 978-0-8151-4387-1.
98. Medicinal and Poisonous Plants of Southern and Eastern Africa - John Mitchell Watt, Maria Gerdina Breyer-Brandwijk (E. & S. Livingstone Ltd. (1962))[page needed]
99. Castor Bean and Jequirity Bean Poisoning at eMedicine
100. "Castor-bean". Canadian Poisonous Plants Information System. Canadian Biodiversity Information Facility. April 11, 2014.
101. Rietschel, Ernst T.; Westphal, Otto (1999). "Endotoxin: Historical Perspectives". In Brade, Helmut. Endotoxin in Health and Disease. CRC. pp. 1–30. ISBN 978-0-8247-1944-9.
102. Thompson, J. N. 2005 The geographic mosaic of coevolution. Chicago, IL: University of Chicago Press.
103. Bronstein, JL. 1994. Our current understand of mutualism. Quarterly Review of Biology 69 (1): 31–51 March 1994
104. Begon, M., J.L. Harper, and C.R. Townsend. 1996. Ecology: individuals, populations, and communities, Third Edition. Blackwell Science Ltd., Cambridge, Massachusetts, USA.
105. Ollerton, J. 2006. "Biological Barter": Interactions of Specialization Compared across Different Mutualisms. pp. 411–435 in: Waser, N.M. & Ollerton, J. (Eds) Plant-Pollinator Interactions: From Specialization to Generalization. University of Chicago Press.
106. van Beneden, Pierre-Joseph (1876). Animal parasites and messmates. London, Henry S. King.

107. Denison RF, Kiers ET 2004. Why are most rhizobia beneficial to their plant hosts, rather than parasitic. Microbes and Infection 6 (13): 1235–1239

108. M.C. Soares; I.M. Côté; S.C. Cardoso & R.Bshary (August 2008). "The cleaning goby mutualism: a system without punishment, partner switching or tactile stimulation". Journal of zoology. 276 (3): 306–312. doi:10.1111/j.1469-7998.2008.00489.x.

109. Porat, D.; Chadwick-Furman, N. E. (2004). "Effects of anemonefish on giant sea anemones: expansion behavior, growth, and survival". Hydrobiologia. 530: 513–520. doi:10.1007/s10750-004-2688-y.

110. Porat, D.; Chadwick-Furman, N. E. (2005). "Effects of anemonefish on giant sea anemones: ammonium uptake, zooxanthella content and tissue regeneration". Mar. Freshw.Behav. Phys. 38: 43–51. doi:10.1080/102362405000_57929.

111. Piper, Ross (2007), Extraordinary Animals: An Encyclopedia of Curious and Unusual Animals, Greenwood Press.

112. Tosh CR, Jackson AL, Ruxton GD (March 2007). "Individuals from different-looking animal species may group together to confuse shared predators: simulations with artificial neural networks". Proc. Biol. Sci. 274 (1611): 827–32. doi:10.1098/rspb.2006.3760. PMC 2093981 . PMID 17251090.

113. Sears CL (October 2005). "A dynamic partnership: celebrating our gut flora". Anaerobe. 11 (5): 247–51. doi:10.1016/j.anaerobe.2005.05.001. PMID 16701579.

114. Rozsa, L; Apari, P. (2012). "Why infest the loved ones – inherent human behaviour indicates former mutualism with head lice" (PDF). Parasitology. 139: 696–700. doi:10.1017/s0031182012000017.

115. "Symbiosis – Symbioses Between Humans And Other Species". Net Industries. Retrieved December 9, 2012.

116. Mt. Pleasant, Jane (2006). "The science behind the Three Sisters mound system: An agronomic assessment of an indigenous agricultural system in the northeast". In John E. Staller; Robert H. Tykot; Bruce F. Benz. Histories of maize: Multidisciplinary approaches to the prehistory, linguistics, biogeography, domestication, and evolution of maize. Amsterdam. pp. 529–537.

117. Gibbon, J Whitfield; foreword by Odum, Eugene P. (2010). Keeping All the Pieces: Perspectives on Natural History and the Environment. Athens, Georgia: University of Georgia Press. pp. 41–42.

118. http://news.discovery.com/animals/whales-dolphins/helpful-dolphins-120502.htm

119. Shipman, Pat (2015). The Invaders: How Humans and Their Dogs Drove Neanderthals to Extinction. Cambridge, Maryland: Harvard University Press.

120. May, R., 1981. Models for Two Interacting Populations. In: May, R.M., Theoretical Ecology. Principles and Applications, 2nd ed. pp. 78–104.

121. García-Algarra, Javier (2014). "Rethinking the logistic approach for population dynamics of mutualistic interactions" (PDF). Journal of Theoretical Biology. 363: 332–343. doi:10.1016/j.jtbi.2014.08.039.

122. Wright, David Hamilton (1989). "A Simple, Stable Model of Mutualism

Incorporating Handling Time". The American Naturalist. 134 (4): 664–667. doi:10.1086/285003.

123. Bascompte, J.; Jordano, P.; Melián, C. J.; Olesen, J. M. (2003). "The nested assembly of plant–animal mutualistic networks". Proceedings of the National Academy of Sciences. 100 (16): 9383–9387. doi:10.1073/pnas.1633576100.

124. Suweis, S.; Simini, F.; Banavar, J; Maritan, A. (2013). "Emergence of structural and dynamical properties of ecological mutualistic networks". Nature. 500: 449–452. doi:10.1038/nature12438.

125. Bastolla, U.; Fortuna, M. A.; Pascual-García, A.; Ferrera, A.; Luque, B.; Bascompte, J. (2009). "The architecture of mutualistic networks minimizes competition and increases biodiversity". Nature. 458 (7241): 1018–1020. doi:10.1038/nature07950.

126. Suweis, S., Grilli, J., Banavar, J. R., Allesina, S., & Maritan, A. (2015) Effect of localization on the stability of mutualistic ecological networks. "Nature communications", 6

127. Lever, J. J.; Nes, E. H.; Scheffer, M.; Bascompte, J. (2014). "The sudden collapse of pollinator communities". Ecology Letters. 17 (3): 350–359. doi:10.1111/ele.12236.

128. Strauss, S.Y. & Zangerl, A.R. (2003). Plant-insect interactions in terrestrial ecosystems. In: Plant-animal interactions: an evolutionary approach (Herrera, C.M. & Pellmyr, O., Eds.). Malden: Blackwell Publishing. pp. 77-106. ISBN 978-0-632-05267-7

129. Wang, B.; Qiu, Y.L. (2006). "Phylogenetic distribution and evolution of mycorrhizae in land plants". Mycorrhiza. 16 (5): 299–363. doi:10.1007/s00572-005-0033-6. PMID 16845554.

130. Lekberg, Y.; Koide, R. T. (2005). "Is plant performance limited by abundance of arbuscular mycorrhizal fungi? A metaanalysis of studies published between 1988 and 2003". New Phytol. 168 (1): 189–204. doi:10.1111/j.1469-8137.2005.01490.x. PMID 16159333.

131. Dighton, J. (2003). Fungi in Ecosystem Processes. New York: Marcel Dekker. ISBN 978-0-8247-4244-7

132. Clay, K. (1990). "Fungal endophytes of grasses". Annu. Rev. Ecol. Syst. 21: 275–297. doi:10.1146/annurev.es.21.110190.001423.

133. Arnold, A.E.; Mejia, L.C.; Kyllo, D.; Rojas, E.I.; Maynard, Z.; Robbins, N.; Herre, E.A. (2003). "Fungal endophytes limit pathogen damage in a tropical tree". Proc. Natl. Acad. Sci. U.S.A. 100 (26): 15649–15654. doi:10.1073/pnas.2533483100. PMC 307622 . PMID 14671327.

134. Thaler J (1999). "Jasmonate-inducible plant defences cause increased parasitism of herbivores". Nature. 399 (6737): 686–688. doi:10.1038/21420.

135. Janzen, D. H. (1966). "Coevolution of mutualism between ants and acacias in Central America". Evolution. 20 (3): 249–275. doi:10.2307/2406628.

136. Vega, F.E.; Posada, F.; Aime, M.C.; Pava-Ripoll, M.; Infante, F.; Rehner, S.A.

(2008). "Entomopathogenic fungal endophytes". Biological Control. 46: 72–82. doi:10.1016/j.biocontrol.2008.01.008.

137. Peterson, R.L.; Massicotte, H.B. & Melville, L.H. (2004). Mycorrhizas: anatomy and cell biology. National Research Council Research Press. ISBN 978-0-660-19087-7.

138. Baverstock, J; Elliot, S.L.; Alderson, P.G.; Pell, J.K. (2005). "Response of the entomopathogenic fungus Pandora neoaphidis to aphid-induced plant volatiles". Journal of Invertebrate Pathology. 89 (2): 157–164. doi:10.1016/j.jip.2005.05.006. PMID 16005016.

139. Gómez-Vidal, S.; Salinas, J.; Tena, M.; Lopez-Llorca, L.V. (2009). "Proteomic analysis of date palm (Phoenix dactylifera L.) responses to endophytic colonization by entomopathogenic fungi". Electrophoresis. 30 (17): 2996–3005. doi:10.1002/elps.200900192. PMID 19676091.

140. Arnold, A.E.; Lutzoni, F. (2007). "Diversity and host range of foliar fungal endophytes: are tropical leaves biodiversity hotspots?". Ecology. 88 (3): 541–549. doi:10.1890/05-1459. PMID 17503580.

141. Leuchtmann, A. (1992). "Systematics, distribution and host specificity of grass endophytes". Nat. Toxins. 1 (3): 150–162. doi:10.1002/nt.2620010303. PMID 1344916.

142. Leuchtmann, A.; Clay, K. (1990). "Isozyme variation in the Acremonium/Epichloe fungal endophyte complex". Phytopathology. 80 (10): 1133–1139. doi:10.1094/Phyto-80-1133.

143. Varma, Ajit; Verma S., Sudha, Nirmal S., Bütehorn B., Franken P. (23 December 1998). "Piriformospora indica, a Cultivable Plant-Growth-Promoting Root Endophyte". Applied and Environmental Microbiology. 65 (6): 2741–2744. PMC 91405 . PMID 10347070. Cite uses deprecated parameter |coauthors= (help)

144. Kogel, Karl-Heinz; Frank Waller,*† Beate Achatz,*†‡Baltruschat H., Fodor J., Becker K., Fischer M., Heier T., Hückelhoven R., Neumann C.,Wettstein D. V.,Franken P. (20 September 2005). "The endophytic fungus Piriformospora indica reprograms barley to salt-stress tolerance, disease resistance, and higher yield". Proc. Natl. Acad. Sci. U.S.A. 102 (38): 13386–13391. doi:10.1073/pnas.0504423102. PMC 1224632 . PMID 16174735. Cite uses deprecated parameter |coauthors= (help)

145. Leuchtmann, A. (1988). "Isozyme relationships of Acremonium endophytes from twelve Festuca species". Mycol. Res. 98: 25–33. doi:10.1016/s0953-7562(09)80331-6.

146. Kluger, C.G.; Dalling, J.W.; Gallery, R.E.; Sanchez, E.; Weeks-Galindo, C.; Arnold, A.E. (2008). "Prevalent host-generalism among fungi associated with the seeds of four neotropical pioneer species". Journal of Tropical Ecology. 24: 332–351. doi:10.1017/s0266467408005026.

147. Leuchtmann, A.; Clay, K. (1989). "Isozyme variation in the fungus Atkinsonella hypoxylon within and among populations of its host grasses". Can. J. Bot. 67 (9): 2600–2607. doi:10.1139/b89-336.

148. Rykard, D.M.; Bacon, C.W.; Luttrell, E.S. (1985). "Host relations of Myriogenospora atramentosa and Balansia epichloe (Clavicipitaceae)". Phytopathology. 75 (8): 950–956. doi:10.1094/Phyto-75-950.

149. Clay, K. (1988). Claviciptaceous fungal endophytes of grasses: coevolution and the change from parasitism to mutualism. In: Coevolution of Fungi with Plants and Animals (Hawksworth, D. & Pirozynski, K., eds.). London: Academic Press.

150. Vicari, M.; Hatcher, P.E.; Ayres, P.G. (2002). "Combined effect of foliar and mycorrhizal endophytes on an insect herbivore". Ecology. 83 (9): 2452–2464. doi:10.1890/0012-9658(2002)083[2452:CEOFAM]2.0.CO;2. ISSN 0012-9658.

151. Klironomos, J. (2008). "Variation in plant response to native and exotic arbuscular mycorrhizal fungi". Ecology. 84 (9): 2292–2301. doi:10.1890/02-0413.

152. Betina, V. (1984). Indole derived tremorgenic toxins. In: Mycotoxins Production, Isolation, Separation and Purification (Betina, V., ed.). Developments in Food Science, Vol. 8. New York: Elsivier.

153. Rutschmann, J. & Stadler, P.A. (1978). Chemical background. In: Ergot Alkaloids and Related Compounds (Berde, B. & Schild, H.O., eds.) Berlin: Springer-Verlag.

154. Rowan, D.D.; Hunt, M.B.; Gaynor, D.L. (1986). "Peramine, a novel insect feeding deterrent from ryegrass infected with the endophyte Acremonium loliae". J. Chem. Soc. Chem. Commun. 1986: 935–936. doi:10.1039/c39860000935.

155. Zhang, D.X.; Nagabhyru, P.; Schardl, C.L. (2009). "Regulation of a Chemical Defense against Herbivory Produced by Symbiotic Fungi in Grass Plants". Plant Physiology. 150 (2): 1072–1082. doi:10.1104/pp.109.138222. PMC 2689992 . PMID 19403726.

156. Popay, A.J.; Prestidge, R.A.; Rowan, D.D. & Dymock, J.J. (1990). The role of Acremonia lolli mycotoxins in insect resistance of perennial ryegrass (Lolium perenne). In: Proc. 1st Int. Symp. Acremonium/Grass interactions (Quisenberry, S.S. & Joost, R.E., eds.) Baton Rouge: Louisiana Agriculture Experiment Station.

157. Clay, K.; Cheplick, G.P. (1989). "Effect of ergot alkaloids from fungal endophyt-infected grasses on fall armyworm (Spodoptera frugiperda)". J. Chem. Ecol. 15: 169–182. doi:10.1007/BF02027781.

158. Patterson, C.G.; Potter, D.A.; Fannin, F.F. (1991). "Feeding deterrency of alkaloids from endophyte-infected grasses to Japanese beetle grubs". Entomol. Exp. Appl. 61 (3): 285–289. doi:10.1111/j.1570-7458.1991.tb01561.x.

159. Prestidge, R.A. & Ball, O.J.-P. (1997). A catch 22: the utilization of endophytic fungi for pest management. In: Multitrophic Interactions in Terrestrial Systems (Gange, A.C. & Brown, V.K., eds.). Oxford: Blackwell Scientific. pp. 171-192. ISBN 978-0-521-83995-2

160. Rowan, D.D.; Dymock, J.J.; Brimble, M.A. (1990). "Effect of fungal metabolite peramine and analogs on feeding and development of Argentine stem weevil (Listronotus bonariensis)". J. Chem. Ecol. 16 (5): 1683–1695. doi:10.1007/BF01014100.

161. Akiyama, K.; Hayashi, H. (2001). "Arbuscular mycorrhizal fungus-promoted

accumulation of two new triterpenoids in cucumber roots". Biosci. Biotechnol. Biochem. 66: 762–769. doi:10.1271/bbb.66.762.

162. Rapparini, F.; Llusia, J.; Penuelas, J. (2008). "Effect of arbuscular mycorrhizal (AM) colonization on terpene emission and content of Artemisia annua L". Plant Biol. 10 (1): 108–122. doi:10.1055/s-2007-964963. PMID 18211551.

163. Smith, S.E. & Read, D.J. (2008). Mycorrhizal symbiosis. (3rd ed.) London: Academic Press. ISBN 978-0-12-370526-6

164. Koricheva, J.; Gange, A.C.; Jones, T. (2009). "Effects of mycorrhizal fungi on insect herbivores: a meta-analysis". Ecology. 90 (8): 2088–2097. doi:10.1890/08-1555.1. PMID 19739371.

165. Gehring, C.; Bennett, A. (2009). "Mycorrhizal fungal-plant-insect interactions: the importance of a community approach". Environ. Entomol. 38 (1): 93–102. doi:10.1603/022.038.0111. PMID 19791601.

166. Langenheim, J.H. (1994). "Higher-plant terpenoids – a phytocentric view of their ecological roles". J. Chem. Ecol. 20 (6): 1223–1280. doi:10.1007/BF02059809.

167. Kagata, H.; Ohgushi, T. (2006). "Bottom-up trophic cascades and material transfer in terrestrial food webs". Ecol. Res. 21: 26–34. doi:10.1007/s11284-005-0124-z.

168. Pineda, A.; Zheng, S.J.; van Loon, J.J.A.; Pieterse, C.M.J.; Dicke, M. (2010). "Helping plants to deal with insects: the role of beneficial soil-borne microbes". Trends in Plant Science. 15 (9): 507–514. doi:10.1016/j.tplants.2010.05.007. PMID 20542720.

169. Gange, A.C.; Brown, V.K.; Aplin, D.M. (2003). "Multitrophic links between arbuscular mycorrhizal fungi and insect parasitoids". Ecol. Lett. 6 (12): 1051–1055. doi:10.1046/j.1461-0248.2003.00540.x.

170. Valenzuela-Soto, J.H.; Estrada-Hernandez, M.G.; Ibarra-Laclette, E.; Delano-Frier, J.P. (2010). "Inoculation of tomato plants (Solanum lycopersicum) with growth-promoting Bacillus subtilis retards whitefly Bemisia tabaci development". Planta. 231 (2): 397–410. doi:10.1007/s00425-009-1061-9. PMID 20041333.

171. Richardson, M.D. (2000). Alkaloids of endophyte-infected grasses: defense chemicals or biological anomalies? In: Microbial Endophytes (Bacon, C.W.; White, J.F., eds.). New York: Marcel Dekker. pp. 323-340. ISBN 0-8247-8831-1

172. Gange, A.C.; Nice, H.E. (1997). "Performance of the thistle gall fly, Urophora cardui, in relation to host plant nitrogen and mycorrhizal colonization". New Phytologist. 137 (2): 335–343. doi:10.1046/j.1469-8137.1997.00813.x.

173. Gosling, P.; Hodge, A.; Goodlass, G.; Bending, G.D. (2006). "Arbuscular mycorrhizal fungi and organic farming". Agric. Ecosyst. Environ. 113: 17–35. doi:10.1016/j.agee.2005.09.009.

174. Latch, G.C.M. (1993). "Physiological interactions of endophytic fungi and their hosts. Biotic stress tolerance imparted to grasses by endophytes". Agric. Ecosyt. Environ. 44: 143–156. doi:10.1016/0167-8809(93)90043-O.

175. Schmidt, S.P.; Osborn, T.G. (1993). "Effects of endophyte-infected tall fescue on animal performance". Agric. Ecosyt. Environ. 44: 233–262. doi:10.1016/0167-

8809(93)90049-U.

176. Latch, G.C.M.; Christensen, M.J.; Gaynor, D.L. (1985). "Aphid detection of endophytic infection in tall fescue". N.Z. J. Agric. Res. 28: 129–132. doi:10.1080/00288233.1985.10427006.

177. Knoch, T.R.; Faeth, S.H. & Arnott, D.S. (1993). Fungal endophytes: plant mutualists via seed predation and germination. Bull. Ecol. Soc. Am. 74(Abstr.): 313.

178. Lacey, L.A.; Neven, L.G. (2006). "The potential of the fungus, Muscodor albus, as a microbial control agent of potato tuber moth (Lepidoptera: Gelechiidae) in stored potatoes". J. Invertebr. Pathol. 91 (3): 195–198. doi:10.1016/j.jip.2006.01.002. PMID 16494898.

179. Madej, C.W.; Clay, K. (1991). "Avian seed preference and weight loss experiments: the role of fungal endophyte-infected tall fescue seeds". Oecologia. 88 (2): 296–302. doi:10.1007/BF00320825.

180. Sadler, K. (1980). Of rabbits and habitat, a long term look. Missouri Conservationist March: 4-7.

181. Mackintosh, C.G.; Orr, M.B.; Gallagher, R.T.; Harvey, I.C. (1982). "Ryegrass staggers in Canadian wapiti deer". N.Z. Vet. J. 36: 106–107.

182. West, C.P.; Izekor, E.; Oosterhuis, D.M.; Robbins, R.T. (1988). "The effect of Acremonium coenophialum on the growth and nematode infestation of tall fescue". Plant Soil. 112: 3–6. doi:10.1007/BF02181745.

183. Clay, K. (1991). Fungal endophytes, grasses, and herbivores. In: Microbial mediation of plant-herbivore interactions (Barbosa, P.; Krischik, V.A. & Jones, C.G., eds.) New York: John Wiley and Sons. pp. 199-226. ISBN 978-0-471-61324-4

184. Newsham, K.K.; Fitter, A.H.; Watkinson, A.R. (1995). "Arbuscular mycorrhizae protect an annual grass from root pathogenic fungi in the field". Journal of Ecology. 83 (6): 991–1000. doi:10.2307/2261180.

185. Gange, A.C. (2000). "Arbuscular mycorrhizal fungi, Collembola and plant growth". Trends in Ecology and Evolution. 15 (9): 369–372. doi:10.1016/S0169-5347(00)01940-6. PMID 10931669.

186. Ahmad, S.; Govindarajan, S.; Funk, C.R.; Johnson-Cicalese, J.M. (1985). "Fatality of house crickets on perennial ryegrass infected with fungal endophyte". Entomol. Exp. Appl. 39 (2): 183–190. doi:10.1111/j.1570-7458.1985.tb03561.x.

187. Cheplick, G.P.; Clay, K. (1988). "Acquired chemical defenses of grasses: the role of fungal endophytes". Oikos. 52 (3): 309–318. doi:10.2307/3565204.

188. Lewis, G.C.; Clements, R.O. (1986). "A survey of ryegrass endophyte (Acremonium loliae) in the U.K. and its apparent ineffectuality on a seedling pest. J. Agric". Sci. 107: 633–638. doi:10.1017/s002185960006980x.

189. Hardy, T.N.; Clay, K.; Hammond, A.M. Jr (1986). "Leaf age and related factors affecting endophyte-mediated resistance to fall armyworm (Lepidoptera: Noctuidae) in tall fescue". Environ. Entomol. 15: 1083–1089.

190. Kindler, S.D.; Breen, J.P.; Springer, T.L. (1991). "Reproduction and damage by Russian wheat aphid (Homoptera:Aphididae) as influenced by fungal endophytes and cool-season turfgrasses". J. Econ. Entomol. 84: 685–692.

191. Koh, S.; Hik, D.S. (2007). "Herbivory mediates grass-endophyte relationships". Ecology. 88 (11): 2752–2757. doi:10.1890/06-1958.1. PMID 18051643.

192. Tibbets, T.M.; Faeth, S.H. (1999). "Neotyphodium endophytes in grasses: deterrents or promoters of herbivory by leaf-cutting ants?". Oecologia. 118 (3): 297–305. doi:10.1007/s004420050730.

193. Clay, K.; Cheplick, G.P.; Marks, S. (1989). "Impact of the fungus Balansia henningsiana on Panicum agrostoides: frequency of infection, plant growth and reproduction, and resistance to pests". Oecologia. 80 (3): 374–380. doi:10.1007/BF00379039.

194. Kruess, A. (2002). "Indirect interaction between a fungal plant pathogen and a herbivorous beetle of the weed Cirsum aravense". Oecologia. 130 (4): 563–569. doi:10.1007/s00442-001-0829-9.

195. Gehring, C.A.; Whitham, T.G. (1994). "Interactions between aboveground herbivores and the mycorrhizal mutualists of plants". Trends in Ecology & Evolution. 9 (7): 251–255. doi:10.1016/0169-5347(94)90290-9.

196. Fitter, A.H.; Garbaye, J. (1994). "Interactions between mycorrhizal fungi and other soil organisms". Plant and Soil. 159 (1): 123–132. doi:10.1007/BF00000101.

197. Gange, A.C.; Brown, V.K.; Aplin, D.M. (2005). "Ecological specificity of arbuscular mycorrhizae: evidence from foliar- and seed-feeding insects". Ecology. 86 (3): 603–611. doi:10.1890/04-0967.

198. Mueller, R.C.; Sthultz, C.M.; Martinez, T.; Gehring, C.A.; Whitham, T.G. (2005). "The relationship between stem-galling wasps and mycorrhizal colonization of Quercus turbinella". Can. J. Bot. 83 (10): 1349–1353. doi:10.1139/b05-105.

199. Faeth, S.H.; Hammon, K.E. (1997a). "Fungal endophytes in oak trees: I. Long-term patterns of abundance and association with leafminers". Ecology. 78 (3): 810–819. doi:10.1890/0012-9658(1997)078[0810:FEIOTL]2.0.CO;2. ISSN 0012-9658.

200. Faeth, S.H.; Hammon, K.E. (1997b). "Fungal endophytes in oak trees: II. Experimental analyses of interactions with leafminers". Ecology. 78 (3): 820–827. doi:10.1890/0012-9658(1997)078[0820:FEIOTE]2.0.CO;2. ISSN 0012-9658.

201. Thomson, J.D. (2003). "When is it mutualism? (An American Society of Naturalists presidential address)". The American Naturalist. 162: S1–S9. doi:10.1086/378683.

202. Arnold, A.E.; Miadlikowskam, J.; Higgins, K.L.; Sarvate, S.D.; Gugger, P.; Way, A.; Hofstetter, V.; Kauff, F.; Lutzoni, F.; et al. (2009). "A phylogenetic estimation of trophic transition networks for ascomycetors fungi: are lichens cradles of symbiotic fungal diversification?". Systematic Biology. 58 (3): 283–297. doi:10.1093/sysbio/syp001. PMID 20525584.

203. Arnold, A.E.; Lamit, L.J.; Gehring, C.A.; Bidartondo, M.I.; Callahan, H. (2010). "Interwoven branches of the plant and fungal trees of life". New Phytologist.

185 (4): 874–878. doi:10.1111/j.1469-8137.2010.03185.x. PMID 20356341.

204. Lane, G.A.; Christenses, M.J. & Miles, C.O. (2000). Coevolution of fungal endophytes with grasses: the significance of secondary metabolites. In: Microbial Endophytes (Bacon, C.W.; White, J.F., eds.). New York: Marcel Dekker. pp. 341-388. ISBN 0-8247-8831-1

205. Bezemer, T.M.; van Dam, N.M. (2005). "Linking aboveground and belowground interactions via induced plant defenses". Trends Ecol. Evol. 20 (11): 617–624. doi:10.1016/j.tree.2005.08.006. PMID 16701445.

206. U'Ren, J.M.; Dalling, J.W.; Gallery, R.E.; Maddison, D.R.; Davis, E.C.; Gibson, C.M.; Arnold, A.E. (2009). "Diversity and evolutionary origins of fungi associated with seeds of a neotropical pioneer tree: a case study for analyzing fungal environmental samples". Mycological Research. 113 (Pt 4): 432–449. doi:10.1016/j.mycres.2008.11.015. PMID 19103288.

207. Clay, K. (1997). Fungal endophytes, herbivores and the structure of grassland communities. In: Multitrophic Interactions in Terrestrial Systems (Gange, A.C. & Brown, V.K., eds.). Oxford: Blackwell Scientific. pp. 151-169. ISBN 978-0-521-83995-2

208. Gange, A.C. (2007). Insect–mycorrhizal interactions: patterns, processes and consequences. In Ecological Communities: Plant Mediation in Indirect Interaction Webs (Ohgushi, T.; Craig, T.P. & Price, P.W., eds.), pp. 124–143. Cambridge, UK: Cambridge University Press. ISBN 978-0-521-85039-1

209. Wilson, D. (1993). "Fungal endophytes: out of sight but should not be out of mind". Oikos. 68 (2): 379–384. doi:10.2307/3544856.

210. Funk, C.R.; Halisky, P.M.; Johnson, M.C.; Siegel, M.R.; Stewart, A.V.; Ahmad, S.; Hurley, R.H.; Harvey, I.C. (1983). "An endophytic fungus and resistance to sod webworms: association in Lolium perenne". Biotechnology. 1 (2): 189–191. doi:10.1038/nbt0483-189.

211. Gallery, R.E.; Dalling, J.W.; Arnold, A.E. (2007). "Diversity, host affinity and distribution of seed-infecting fungi: a case-study with neotropical Cecropia". Ecology. 88: 582–588. doi:10.1890/05-1207.

212. Dalling, J.W.; Davis, A.S.; Schutte, B.J.; Arnold, A.E. (2011). "Seed survival in soil: interacting effects of predation, dormancy and the soil microbial community". Journal of Ecology. 99 (1): 89–95. doi:10.1111/j.1365-2745.2010.01739.x.

213. Van WH, der Putten W (2003). "Plant defense belowground and spatiotemporal processes in natural vegetation". Ecology. 84 (9): 2269–2280. doi:10.1890/02-0284.

214. Wardle, D.A.; Bardgett, R.D.; Klironomos, J.N.; Setala, H.; Van, WH; der Putten, W.H.; Wall, D.H. (2004). "Ecological linkages between aboveground and belowground biota". Science. 304 (5677): 1629–1633. doi:10.1126/science.1094875. PMID 15192218.

215. Molinari, N.; Knight, C. (2010). "Correlated evolution of defensive and nutritional traits in native and non-native plants". Bot. J. Linn. Soc. 163 (1): 1–13. doi:10.1111/j.1095-8339.2010.01050.x.

216. Dighton, J.; Jansen, A.E. (1991). "Atomospheric pollutants and ectomycorrhizas: more questions than answers?". Environ. Pollut. 73 (3–4): 179–204. doi:10.1016/0269-7491(91)90049-3. PMID 15092077.

217. Marks, S.; Clay, K. (1990). "Effects of CO2 enrichment, nutrient addition and fungal endophyte-infection on the growth of two grasses". Oecologia. 84: 207–214. doi:10.1007/BF00318273.

218. Currano, E.D.; Wilf, P.; Wing, S.L.; Labandeira, C.C.; Lovelock, E.; Royer, D. (2008). "Sharply increased insect herbivory during the Paleocene-Eocene Thermal Maximum". Proceedings of the National Academy of Sciences of the United States of America. 105 (6): 1960–1964. doi:10.1073/pnas.0708646105. PMC 2538865 . PMID 18268338.

219. Marks, S.; Lincoln, D.E. (1996). "Antiherbivore defense mutualism under elevated carbon dioxide levels: a fungal endophyte and grass". Environ. Entomol. 25 (3): 618–623.

220. Clay, K. (1994). The potential role of endophytes in ecosystems. In: Biotechnology of endophytic fungi of grasses. Boca Raton: CRC Press. pp. 73-86. ISBN 978-0-8493-6276-7

221. Bailey, V. (1903). "Sleepy grass and its effect on horses". Science. 17 (427): 392–393. doi:10.1126/science.17.427.392. PMID 17735119.

222. Nobindro, U. (1934). "Grass poisoning among cattle and goats in Assam". Indian Vet. J. 10: 235–236.

223. Siegel, M.R.; Latch, G.C.M.; Johnson, M.C. (1985). "Acremonium fungal endophytes of tall fescue and perennial ryegrass: significance and control". Plant Dis. 69 (2): 179–183.

224. Clay, K. (1988). "Fungal endophytes of grasses – a defensive mutualism between plants and fungi". Ecology. 69 (1): 10–16. doi:10.2307/1943155.

225. Hartley, S.E.; Gange, A.C. (2009). "Impacts of Plant Symbiotic Fungi on Insect Herbivores: Mutualism in a Multitrophic Context". Annu. Rev. Entomol. 54: 323–342. doi:10.1146/annurev.ento.54.110807.090614.

226. Fontana, A.; Reichelt, M.; Hempel, S.; Gershenzon, J.; Unsicker, S.B. (2009). "The effects of arbuscular mycorrhizal fungi on direct and indirect defense metabolites of Plantago lanceolata L". J. Chem. Ecol. 35 (7): 833–843. doi:10.1007/s10886-009-9654-0. PMC 2712616 . PMID 19568812.

227. Bultman, T.L.; Bell, G.; Martin, W.D. (2004). "A fungal endophyte mediates reversal of wound-induced resistance and constrains tolerance in a grass". Ecology. 85 (3): 679–685. doi:10.1890/03-0073.

228. Pieterse, C.M.J.; Dicke, M. (2007). "Plant interactions with microbes and insects: from molecular mechanisms to ecology". Trends Plant Sci. 12 (12): 564–569. doi:10.1016/j.tplants.2007.09.004. PMID 17997347.

229. Zheng, S.J.; Dicke, M. (2008). "Ecological genomics of plant-insect interactions: from gene to community". Plant Physiol. 146 (3): 812–817. doi:10.1104/pp.107.111542. PMC 2259077 . PMID 18316634.

230. Simmons, L.; Bultman, T.L.; Sullivan, T.J. (2008). "Effects of Methyl Jasmonate and an Endophytic Fungus on Plant Resistance to Insect Herbivores". J. Chem. Ecol. 34 (12): 1511–1517. doi:10.1007/s10886-008-9551-y. PMID 18925382.

231. Wink, M. (2008). "Plant secondary metabolism: Diversity, function and its evolution". Natural Product Communications. 3 (8): 1205–1216.

232. Lambais, M.R. (2001). "In silico differential display of defense-related expressed sequence tags from sugarcane tissues infected with diazotrophic endophytes". Genetics and Molecular Biology. 24 (1): 103–111. doi:10.1590/S1415-47572001000100015.

233. Shrivastava, G.; Rogers, M.; Wszelaki, A.; Panthee, D.R.; Chen, F. (2010). "Plant Volatiles-based Insect Pest Management in Organic Farming". Plant Sciences. 29 (2): 123–133. doi:10.1080/07352681003617483.

234. Hoveland, C.S. (1993). "Importance and economic significance of the Acremonium endophytes to performance of animals and grass plant". Agric. Ecosyst. & Environ. 44 (1): 3–12. doi:10.1016/0167-8809(93)90036-O.

235. Jarvis, B.B.; Wells, K.M.; Lee, Y.W.; Bean, G.A.; Kommendahl, T.; Barros, C.S. L.; Barros, S.S. (1987). "Macrocyclic triocothecene mycotoxins in Brazilian species of Baccharis". Phytopathology. 12: 111–128.

236. Steudemann, J.A.; Hoveland, C.S. (1988). "Fescue endophyte: history and impact on animal agricultural". J. Product Agric. 1: 39–44.

237. West, C.P. & Gwinn, K.D. (1993). Role of Acremonium in drought, pest and disease tolerance of grasses. In: Proc. 2nd Int. Symp. Acremonium/grass interactions: plenary papers. (Hume, D.E.; Latch, G.C.M. & Easton, H.S., eds.) Palmerston North, NZ: AgResearch, Grasslanda Research Centre.

238. Rabin, L.B.; Pacovsky, R.S. (1985). "Reduced larva growth of two Lepidoptera (Noctuidae) on excised leaves of soybean infected with amycorrhizal fungus". J. Econ. Entom. 78: 1358–1363.

239. Gange, A.C.; West, H.M. (1994). "Interactions between arbuscular mycorrhizal fungi and foliar-feeding insects in Plantago lanceolata L". New Phytol. 128: 79–87. doi:10.1111/j.1469-8137.1994.tb03989.x.

240. Akello, J.; Dubois, T.; Coyne, D.; Kyamanywa, S. (2008). "Endophytic Beauveria bassiana in banana (Musa spp.) reduces banana weevil (Cosmopolites sordidus) fitness and damage". Crop Protection. 27 (11): 1437–1441. doi:10.1016/j.cropro.2008.07.003.

241. Jallow, M.F.A.; Dugassa-Gobena, D.; Vidal, S. (2004). "Indirect interaction between and unspecialized endophytic fungus and a polyphagous moth". Basic and Applied Ecology. 5 (2): 183–191. doi:10.1078/1439-1791-00224.

242. Dent, D. (1991) Insect pest management. Wallingford, UK: CAB International. ISBN 0-85199-340-0

243. Murray, F.R.; Latch, G.M.C.; Scott, D.B. (1992). "Surrogate transformation of perennial ryegrass, Lolium perenne, using genetically modified Acremonium endophyte". Mol. Gen, Genet. 233: 1–9. doi:10.1007/BF00587554.

244. Shennan, C. (2008). "Biotic interactions, ecological knowledge and agriculture". Phil. Trans. R. Soc. B. 363 (1492): 717–739. doi:10.1098/rstb.2007.2180. PMC 2610106 . PMID 17761466.

245. Black, M.H. & Halmer, P. (2006). The encyclopedia of seeds: science, technology and uses. Wallingford, UK: CABI. pp. 226. ISBN 978-0-85199-723-0.

246. Schade, R.; Andersohn, F.; Suissa, S.; Haverkamp, W.; Garbe, E. (2007). "Dopamine agonists and the risk of cardiac-valve regurgitation". New England Journal of Medicine. 356 (1): 29–38. doi:10.1056/NEJMoa062222. PMID 17202453.

247. Correia, T.; Grammel, N.; Ortel, I.; Keller, U.; Tudzynski, P. (2001). "Molecular cloning and analysis of the ergopeptine assembly system in the ergot fungus Claviceps purpurea". Chem Biol. 10 (12): 1281–1292. doi:10.1016/j.chembiol.2003.11.013. PMID 14700635.

248. Gray, W.D. (1959). The Relation of Fungi to Human Affairs. New York: Henry Holt and Company, Inc.

CHAPTER 6

General Characteristics of Viroids

Chapter Contents

Viroids are small RNA particles which infect plants. Viroids are remarkable for their extremely tiny size (only a few hundred nucleic acid bases, the smallest with only 220) relative to the larger viruses, which contain at least 2,000 bases. Viroids were discovered and given their name by Theodor Diener, a plant pathologist at the Agricultural Research Service in Maryland, in 1971. Viroids are a subviral agent about 80 times smaller than typical viruses. Their discovery was considered a major breakthrough in 20th century biology.

INTRODUCTION

Viroids are the smallest infectious pathogens known, larger only than prions, which, as nucleic acid-free, misfolded protein molecules, only whose abnormal conformation is transmitted, are not strictly infectious pathogens. Viroids consist solely of short strands of circular, single-stranded RNA without protein coats. They are mostly plant pathogens, some of which are of economic importance. Viroid genomes are extremely small in size, ranging from 246 to 467 nucleobases. In comparison, the genome of the smallest known viruses capable of causing an infection by themselves are around 2,000 nucleobases in size. The human pathogen hepatitis D virus is a "defective" RNA virus similar to a viroid.

Viroids, the first known representatives of a new domain of "sub-viral pathogens", were discovered, initially characterized, and named by Theodor Otto Diener, plant pathologist at the U.S Department of Agriculture's Research Center in Beltsville, Maryland, in 1971. The first viroid to be identified was Potato spindle tuber viroid (PSTVd). Some 33 viroid species have been identified.

Viroids do not code for any protein. A viroid's replication mechanism uses RNA polymerase II, a host cell enzyme normally associated with synthesis of messenger RNA from DNA, which instead catalyzes "rolling circle" synthesis of new RNA using the viroid's RNA as a template. Some viroids are ribozymes, having catalytic properties which allow self-cleavage and ligation of unit-size genomes from larger replication intermediates.

With Diener's 1989 hypothesis that viroids may represent "living relics" from the widely assumed, ancient, and non-cellular RNA world—extant before the evolution of DNA or proteins—viroids have assumed significance beyond plant pathology to evolutionary science, by representing the most plausible RNAs capable of performing crucial steps in abiogenesis, the evolution of life from inanimate matter.

Taxonomy

Family Pospiviroidae

Genus Pospiviroid; type species: Potato spindle tuber viroid ; 356–361 nucleotides(nt)

Genus Pospiviroid; type species: Citrus exocortis ; 368–467 nt

Genus Hostuviroid; type species: Hop stunt viroid ; 294–303 nt

Genus Cocadviroid; type species: Coconut cadang-cadang viroid; 246–247 nt

Genus Apscaviroid; type species: Apple scar skin viroid ; 329–334 nt

Genus Coleviroid; type species: Coleus blumei viroid 1 ; 248–251 nt

Potato Spindle Tuber Viroid

The Potato spindle tuber viroid ("PSTVd") was the first viroid to be identified PSTVd is a small, circular RNA molecule closely related to the Chrysanthemum stunt viroid. Present within the viroidal RNA is the Pospiviroid RY motif stem loop. The natural hosts are potatoes (Solanum tuberosum) and tomatoes (Solanum lycopersicum). All potatoes and tomatoes are susceptible to PSTVd and there is no form of natural resistance. Natural infections have also been seen in avocados and infections in other solanaceous crops have been induced in the laboratory.

Strains and their Symptoms

Different strains of PSTVd exist and symptoms range from mild to severe. Mild strains produce no obvious symptoms. Symptoms in severe strains are dependent on environmental conditions and are most severe in hot conditions. Symptoms may be mild in initial infections but become progressively worse in the following generations. Common symptoms of severe infections include colour changes in the foliage, smaller leaves and spindle like elongation. Sprouting also occurs at a slower rate than in unaffected potatoes. Infected tomatoes are slower to show symptoms which include stunted growth with a 'bunchy top' caused by shortened internodes. Leaves become yellow or purple and often become curled and twisted. Necrosis eventually occurs in the veins of the bottom and middle leaves and the top leaves don't die but are reduced in size. Fruit ripening is also affected leading to hard, small, dark green tomatoes.

Long distance spread of PSTVd usually occurs via infected seeds but transmission via aphids (Myzus persicae) also occurs but only in the presence of PLRV (potato leaf roll virus). Mechanical transmission also occurs once it has been introduced to an area.

Primary and Secondary Structure of PSTVD

PSTVd comprises 359 nucleotides.

Primary Structure

1 CGGAACUAAA CUCGUGGUUC CUGUGGUUCA CACCUGACCU CCUGAGCAGA AAAGAAAAAA
61 GAAGGCGGCU CGGAGGAGCG CUUCAGGGAU CCCCGGG**GAA AC**CUGGAGCG AACUGGCAAA
121 AAAGGACGGU GGGGAGUGCC CAGCGGCCGA CAGGAGUAAU UCCCGCCGAA ACAGGGUUUU
181 CACCCUUCCU UUCUUCGGGU GUCCUUCCUC GCGCCCGCAG GACCACCCCU CGCCCCCUUU
241 GCGCUGUCGC UUCGGCUACU ACCCGGUGGA AACAACUGAA GCUCCCGAGA ACCGCUUUUU
301 CUCUAUCUUA CUUGCUUCGG GGCGAGGGUG UUUAGCCCUU GGAACCGCAG UUGGUUCCU

Secondary Structure

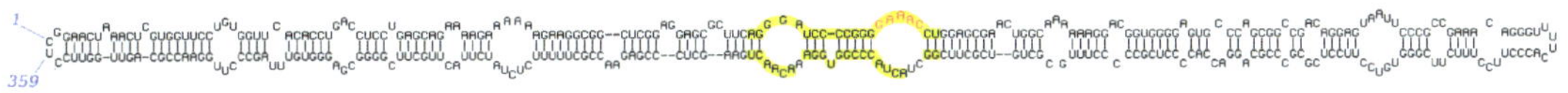

Cadang-Cadang

Cadang-cadang is a disease caused by Coconut cadang-cadang viroid (CCCVd), a lethal viroid of Coconut (Cocos nucifera), Anahaw (Livistona rotundifolia) Buri (Corypha utan), and African Oil Palm (Elaeis guineensis). The name cadang-cadang comes from the word gadang-gadang that means dying in Bicol. It was originally reported on San Miguel Island in the Philippines in 1927/1928. "By 1962, all but 100 of 250,000 palms on this island had died from the disease," indicating an epidemic. Every year one million coconut palms are killed by CCCVd and over 30 million coconut palms have been killed since Cadang-cadang has been discovered. CCCVd directly affects the production of copra, a raw material for coconut oil and animal feed. Total losses of about 30 million palms and annual yield losses of about 22,000 tons of copra have been attributed to Cadang-cadang disease in the Philippines.

Characteristics

Viroids are small, single-stranded RNA molecules, ranging from 246 to 375 nucleotides long. Unlike viruses, they do not code for protein coats but contain genes for autonomous replication. With abilities to cause serious disease, they are more commonly found in a latent stage, and their mode of infection is mainly mechanical, though documented cases exist of vertical transmission through pollen and seed. The sequence and structure prediction of CCCVd has been documented. There are four low-molecular-weight RNA species (ccRNAs) found associated with cadang-cadang disease. Of the four, two are related with the early stage of the disease: ccRNA 1 fast and ccRNA 2 fast. After several years, two other species appear and predominate: ccRNA 1 slow and ccRNA 2 slow. Moreover, they share sequence homology with other viroids.

Conditions for a viroid to infect its host include wounds on the host or infected pollen grain deposited into an ovule. Not all of the mentioned conditions pertain to cadang-cadang.

Tinangaja disease is caused by coconut trinangaja viroid (CTiVd), which has 64% sequence homology with CCCVd. This disease has been found in Guam. Coconuts from Asia and South Pacific have been found to have viroids with similar nucleic acid sequences of CCCVd. The pathogenicity of CTiVd is uncertain.

Coconut cadang-cadang viroid, also known as CCCVd, is responsible for a lethal disease of coconut plant first reported in the early 1930 in the Philippines. CCCVd is the smallest known pathogen and it is biologically distinct from other viroids; it consists of circular or lineal single-stranded RNA with a basic size of 246 or 247, it is thought it can be transmitted by seed or pollen (with low transmission rates) and occur in almost all plant parts. Once infected, coconut palm shows yellow leaf spots and nut production ceases; from appearance of first symptoms to tree death, time ranges from around 8 to 16 years, but the pest is generally greater in older plants. It is estimated that over 30 million coconut palms have been killed by cadang-cadang since it was first recognised and the loss of production is valued at about \$80 – \$100 for each planting site occupied by an infected tree.

Molecular Structure

CCCVd are small, circular, and single-stranded. They cannot replicate themselves, and therefore are completely dependent on a host.

CCCVd have a sequence of 246 nucleotides, 44 of which are common with most viroids. CCCVd can add a cytosine residue in the 197 position and increase the size to 247 nucleotides. It is known that forms of CCCVd with 247 nucleotides cause severe symptoms. Their minimum size is 246, but they can produce forms from 287 to 301 nucleotides.

In CCCVd, changes in molecular forms are related to the steps of the disease. There are four different RNAs found in CCCVd, two of them fast and the other two slow (the difference between fast and slow RNA is their mobilities in electrophoresis gel). In the early stages of the disease, RNA fast 1 and RNA fast 2 appear, while in the late stages RNA slow 1 and RNA slow 2 are detected.

Geographical Distribution

The CCCVd is widely spread in Philippines and it is mostly found in Bicol Region, Masbate, Catanduanes, Samar and other smaller islands in the zone. It is known that the present northernmost boundary is at the latitude of Manila and the southernmost at the latitude of Homonhon Island. This fact is important due to the proximity of the disease to the major coconut and oil palm growing area of Mindanao. An isolated focus of infection has been found on Solomon Islands, Oceania. The available data at the present suggests that the epidemiology of cadang-cadang viroid may not be spreading into any specific route.

Variants

In Guam, a similar viroid known as coconut tinangaja viroid (CTiVd) has been found that causes a similar disease named tinangaja disease. This viroid has 64% sequence homology with the cadang-cadang viroid. There are other related viroids with the CCCVd, which are found in Asia and the South Pacific. They have a high degree of homology but the pathogenicity is uncertain.

Transmission

The mode of natural inoculation of the viroid is unknown. There has been evidence that the transmission through pollen and seed can occur but they have a very low transmission rates: progenies of healthy palms pollinated with diseased pollen, exhibited disease symptoms 6 years after germination.

CCCVd can spread through mechanical inoculation primarily through contaminated farms tools such as harvesting scythes or machetes, due to the improper sanitary conditions. The efficiency of the mechanical inoculation has been influenced by factors such as the age of the test plant and the mode of inoculation.

Experimental transmission of the CCCVd using insects as vector has been unsuccessful although there are few evidences that the transmission of the viroid might be related to the wounds of coleopteran insects, but this has not been properly investigated.

The natural hosts of cadang-cadang identified are *Cocos nucifera, Corypha elata, Elaeis guineensis* and *Oredoxa regia*. The experimental hosts are *Adonidia merrillii, Areca catechu, Caryota cumingii, Chrysalidocarpus lutescens, Livistona rotundifolia, Phoenix dactylifera, Ptychosperma macarthurii* and *Roystonea regia*.

Detection

The diagnosis by symptoms is not reliable enough so it's better to do a moleculardiagnosis based in test samples. Some of these methods (like Dot-blot hybridisation) allow the scientists to detect the viroid even six months before noticing the first symptoms.

The first step is the purification to obtain the nucleic acids of the plant cells. The leaves of the plant located four or more below the spare leaf are cut. Afterwards, they are blended (homogenize) with Na2 So3. Then the extract is filtered and clarified by centrifugation (10.000 g during 10 minutes). The next step is to add polyethylene glycol (PEG). Finally, after nearly two hours of incubation at 4 °C, and after another centrifugation (at low speed) the nucleic acids can be extracted by chloroform procedures, for example.

When approximately 1 g of coconut tissue has been purified, the electrophoresis method can be started, which will help to identify the viroid by its relative mobility. The CCCVd is analysed in one or two dimensional polyacrylamide gels with a silver stain.

The viroid can also be detected by a more sensitive method called Dot blot molecular hybridization. In this method CCCVd is amplified by the PCR (polymerase chain reaction) and the clones of CCCVd are used as templates to synthesize a complementary DNA or RNA chain. These sequences are radioactively labelled so when they are put over the samples with the intention to analyse (on a supporting membrane) and exposed to x-ray film, then if CCCVd is present it will appear as a dark colour. This dark tonality only appears when nucleic acid hybridization occurs.

Pharmacology and Treatment

Coconut cadang-cadang disease has no treatment yet. However, chemotherapy with antibiotics has been tried with tetracycline solutions; antibiotics failed trying to alter progress of the disease since they had no significant effect on any of the studied parameters. When the treated plants were at the early stage, tetracycline injections failed to prevent the progression of the palms to more advanced stages, nor did they affect significantly the mean number of spathes or nuts. Penicillin treatment had no apparent improvement either.

Control strategies are elimination of reservoir species, vector control, mild strain protection and breeding for host resistance. Eradication of diseased plants is usually performed to minimize spread but is of dubious efficacy due to the difficulties of early diagnosis as the virus etiology remains unknown and the one discovered are the three main stages in the disease development.

Easily Confused Pests

Coconut Cadang Cadang Viroid (CCCVd) can be confused with another viroid called Coconut Tinangaja Viroid (CTiVd). That is because both viroids have 64% of their sequence of nucleotides in common. They both affect coconut palms (monocots), and infected plants have similar symptoms: spots on the leaves reduced top, yellow palms and even death.

There are few differences between both viroids in the effects caused on fruit: CTiVd causes small nuts, while CCCVd causes round and reduced nuts. Moreover, CTiVd has 254 nucleotides of length, while CCCVd has 246.

Symptoms

Coconut

Symptoms of cadang-cadang develop slowly over 8 to 15 years making it difficult to diagnose at an early time. There are three main "stages" of defined series of characteristics: early, medium, and late stages. The first symptoms in the early stage develop within two to four years of infection. These symptoms include scarification of the coconuts which also become rounded. The leaves (fronds) display bright yellow spots. About two years later, during the medium stage, the inflorescences become stunted and eventually killed, so no more coconuts are produced. Yellow spots are larger and in greater abundance to give the appearance of chlorosis. During the final stage, roughly 6 years after the first symptoms are recorded, the yellow/bronze fronds start to decrease in size and number. Finally, all the leaves coalesce, leaving just the trunk of the palm "standing like a telephone pole". Palms under 10 years of age are rarely affected by cadang-cadang; the incidence of disease increases until about 40 years of age and then plateaus. "No recovery has ever been observed, and the disease is always fatal". African oil palm has similar symptoms as coconut but also have orange spotting on palms.

Environment

Geographical Location

Philippines: Southern Luzon, Samar, Masbate and other small islands. Tinangaja was found in Guam on Marianas Island.Evidence shows that CCCVd is negatively correlated with altitude.

Dispersal

The CCCVd viroid can be transmitted by pollen, seed, and mechanical transmission; it infects almost all parts of the host after flowering.

Insect Vectors

No insect vector is known at this time, and rate of dispersal is not related to proximity of clusters.

Control

No definite control measures exist at the present.

- Genetic resistance and Vector control are not options because resistant/tolerant varieties have yet to be discovered, and there is no known vector of CCCVd.

- Eradication is ineffective because of the long latent period between infection and appearance of symptoms, which is approximately 1 to 2 years.
- Cross-protection (see also Influenza vaccine: Cross-protection) is a possibility for the future. Cross infection means inoculating the coconut with a mild strain of the viroid to give the coconut tree some degree of protection from infection by the killer form. This is similar to Edward Jenner›s pioneering smallpox vaccine. He used cowpox to confer induced immunity on humans. But in the case of Cadang-cadang, this is still under research.

Family Avsunviroidae

- Genus *Avsunviroid*; type species: *Avocado sunblotch viroid* ; 246–251 nt
- Genus *Pelamoviroid*; type species: *Peach latent mosaic viroid* ;335–351 nt
- Genus *Elaviroid*; type species: *Eggplant latent viroid* ; 332–335 nt

AVSUNVIROIDAE

The Avsunviroidae are a family of viroids. At present three members are known. They consist of RNA genomes between 246-375 nucleotides in length. They are single-stranded covalent circles and have intramolecular base pairing. All members lack a central conserved region.

Replication

Replication occurs in the chloroplasts of plant cells. Key features of replication include no helper virus required and no proteins are encoded for. Unlike the other family of viroids, Pospiviroidae, Avsunviroidae are thought to replicate via a symmetrical rolling mechanism. It is thought the positive RNA strand acts as a template to form negative strands with the help of an enzyme thought to be RNA polymerse plus 3 II. The negative RNA strands are then cleaved by ribozyme activity and circularises. A second rolling circle mechanism forms a positive strand which is also cleaved by ribozyme activity and then ligated to become circular. The site of replication is unknown but it is thought to be in the chloroplast and in the presence of Mg^{2+} ions.

AVOCADO SUNBLOTCH VIROID

Avocado sunblotch viroid (ASBV) is an important disease affecting avocado trees.

Infections result in lower yields and poorer quality fruit. ASBV is the smallest known viroid that infects plants and is transmitted by pollen and infected seeds or budwood.

Trees infected with the viroid often show no symptoms other than a reduction in yield, (by 30% or more). However, they are still carriers and can pass the disease onto other plants. Symptoms in more serious infections include depressed longitudinal streaks of yellow in the fruit. The fruit may also become red or white in colour. Symptoms in the leaf are uncommon but include bleached veins and petioles. Rectangular cracking patterns also occur in the bark of older branches.

The amount (titre) of viroid particles present in avocado trees varies a great deal. Viroids levels can vary by 1000 times between branches on the same tree and by 10000 times between trees. Infected but symptomless trees have a higher concentration of viroid particles than those showing symptoms. Symptomless trees also represent a greater danger in terms of spread of the viroid.

PCR is used to detect infection.

PEACH LATENT MOSAIC VIROID

Peach latent mosaic viroid is a type species from the genus Pelamoviroid, which belongs to the family Avsunviroidae. This family is characterized as having chloroplastic viroids with hammerhead ribozymes. Peach latent mosaic viroid is a 336-351nt circular RNA which has a branched formation. This branched formation is stabilised by a pseudoknot between two kissing loops. Peach latent mosaic viroid was first described in the 1980s in Spain by a group of scientists.

It is present in all peach- and nectarine-producing areas of the world including Europe, Asia, North America and South America and the frequency of naturally occurring infection is high.

Before the development of symptoms the disease is latent in peach trees for approximately 5–7 years. The symptoms of the disease include necrosis of buds, delayed shoot development, necrotic branches, premature ageing of trees, flower streaking, ripening deformations, enlarged rounded stones, circular discoloured areas on the fruit skin and in some cases mosaic, blotch, vein banding or calico appearance on infected leaves. Peach latent mosaic viroid is horizontally spread from plant to plant across a field by propagation of infected buds, pruning tools and green peach aphids. Vertical transmission of peach latent mosaic viroid through seeds cannot occur.

Transmission

The reproduction mechanism of a typical viroid. Leaf contact transmits the viroid. The viroid enters the cell via its plasmodesmata. RNA polymerase II catalyzes rolling-circle synthesis of new viroids.

Viroid infections are transmitted by cross contamination following mechanical damage to plants as a result of horticultural or agricultural practices. Some are transmitted by aphids, and they can also be transferred from plant to plant by leaf contact.

Replication

Viroids replicate in the nucleus (*Pospiviroidae*) or chloroplasts (*Avsunviroidae*) of plant cells in three steps through an RNA-based mechanism. They require RNA polymerase II, a host cell enzyme normally associated with synthesis of messenger RNA from DNA, which instead catalyzes "rolling circle" synthesis of new RNA using the viroid as template

RNA Silencing

There has long been uncertainty over how viroids induce symptoms in plants without encoding any protein products within their sequences. Evidence suggests that RNA silencing is involved in the process. First, changes to the viroid genome can dramatically alter its virulence. This reflects the fact that any siRNAs produced would have less complementary base pairing with target messenger RNA. Secondly, siRNAs corresponding to sequences from viroid genomes have been isolated from infected plants. Finally, transgenic expression of the noninfectious hpRNA of potato spindle tuber viroid develops all the corresponding viroid-like symptoms. This indicates that when viroids replicate via a double stranded intermediate RNA, they are targeted by a dicer enzyme and cleaved into siRNAs that are then loaded onto the RNA-induced silencing complex. The viroid siRNAs contain sequences capable of complementary base pairing with the plant›s own messenger RNAs, and induction of degradation or inhibition of translation causes the classic viroid symptoms.

Living Relics of the RNA World

Diener's 1989 hypothesis proposed that unique properties of viroids make them more plausible macromolecules than introns, or other RNAs considered in the past as possible "living relics" of a hypothetical, pre-cellular RNA world. If so, viroids have assumed significance beyond plant virology for evolutionary science, because their properties make them more plausible candidates than other RNAs to perform crucial steps in the evolution of life from inanimate matter (abiogenesis). These properties are:

- viroids' small size, imposed by error-prone replication
- their high guanine and cytosine content, which increases stability and replication fidelity
- their circular structure, which assures complete replication without genomic tags
- existence of structural periodicity, which permits modular assembly into enlarged genomes
- their lack of protein-coding ability, consistent with a ribosome-free habitat
- replication mediated in some by ribozymes—the fingerprint of the RNA world

Diener›s hypothesis was mostly forgotten until 2014, when it was resurrected in a review article by Flores et al., in which the authors summarized Diener›s evidence supporting his hypothesis (see above). In the same year, *New York Times* science writer Carl Zimmer published a popularized piece that mistakenly credited Flores et al. with the hypothesis' original conception.

The presence, in extant cells, of RNAs with molecular properties predicted for RNAs of the RNA World constitutes another powerful argument supporting the RNA World hypothesis.

In the 1920s, symptoms of a previously unknown potato disease were noticed in New York and New Jersey fields. Because tubers on affected plants become elongated and misshapen, they named it the potato spindle tuber disease.

The symptoms appeared on plants onto which pieces from affected plants had been budded—indicating that the disease was caused by a transmissible pathogenic agent. However, a fungus or bacterium could not be found consistently associated with symptom-bearing plants, and therefore, it was assumed the disease was caused by a virus. Despite numerous attempts over the years to isolate and purify the assumed virus, using increasingly sophisticated methods, these were unsuccessful when applied to extracts from potato spindle tuber disease-afflicted plants.

In 1971 Theodor O. Diener showed that the agent was not a virus, but a totally unexpected novel type of pathogen, one-80th the size of typical viruses, for which he proposed the term "viroid". Parallel to agriculture-directed studies, more basic scientific research elucidated many of viroids› physical, chemical, and macromolecular properties. Viroids were shown to consist of short stretches (a few hundred nucleobases) of single-stranded RNA and, unlike viruses, did not have a protein coat. Compared with other infectious plant pathogens, viroids are extremely small in size, ranging from 246 to 467 nucleobases; they thus consist of fewer than 10,000 atoms. In comparison, the genomes of the smallest known viruses capable of causing an infection by themselves are around 2,000 nucleobases long.

In 1976, Sänger et al. presented evidence that potato spindle tuber viroid is a "single-stranded, covalently closed, circular RNA molecule, existing as a highly base-paired rod-like structure"—believed to be the first such molecule described. Circular RNA, unlike linear RNA, forms a covalently closed continuous loop, in which the 3› and 5› ends present in linear RNA molecules have been joined together. Sänger et al. also provided evidence for the true circularity of viroids by finding that the RNA could not be phosphorylated at the 5› terminus. Then, in other tests, they failed to find even one free 3› end, which ruled out the possibility of the molecule having two 3' ends. Viroids thus are true circular RNAs.

The single-strandedness and circularity of viroids was confirmed by electron microscopy, and Gross et al. determined the complete nucleotide sequence of potato spindle tuber viroid in 1978. PSTVd was the first pathogen of a eukaryotic organism for which the complete molecular structure has been established. Over thirty plant diseases have since been identified as viroid-, not virus-caused, as had been assumed.

REFERENCES

1. Flores Ricardo; Delgado Sonia; Rodio María-Elena; Ambrós Silvia; Hernández Carmen; di Serio Francesco (2006). "Peach latent mosaic viroid: not so latent". Molecular Plant Pathology. 7 (4): 209–221(13). doi:10.1111/j.1364-3703.2006.00332.x. PMID 20507441.

2. "PEACH LATENT MOSAIC VIROID DECLARATION". Retrieved 2007-06-05.
3. Hadidi, A., Giunchedi, L., Shamloul, A. M., Poggi-Pollini, C., and Amer, M. A. (1997). "Occurrence of peach latent mosaic viroid in stone fruits and its transmission with contaminated blades" (PDF). Plant Dis. 81 (2): 154–158. doi:10.1094/PDIS.1997.81.2.154.
4. Lewin, Benjamin.; Krebs, Jocelyn E.; Kilpatrick, Stephen T.; Goldstein, Elliott S.; Lewin, Benjamin. Genes IX. (2011). Lewin's genes. Sudbury, Mass.: Jones and Bartlett. p. 23. ISBN 9780763766320.
5. Alves C, Branco C, Cunha C (2013). "Hepatitis delta virus: a peculiar virus". Adv Virol. 2013: 560105. doi:10.1155/2013/560105. PMC 3807834 . PMID 24198831.
6. Diener TO (August 1971). "Potato spindle tuber "virus". IV. A replicating, low molecular weight RNA". Virology. 45 (2): 411–28. doi:10.1016/0042-6822(71)90342-4. PMID 5095900.
7. "ARS Research Timeline – Tracking the Elusive Viroid". 2006-03-02. Retrieved 2007-07-18.
8. Tsagris EM, Martínez de Alba AE, Gozmanova M, Kalantidis K (September 2008). "Viroids". Cell. Microbiol. 10 (11): 2168–79. doi:10.1111/j.1462-5822.2008.01231.x. PMID 18764915.
9. Flores, Ricardo; Di Serio, Francesco; Hernández, Carmen (February 1997). "Viroids: The Noncoding Genomes". Seminars in Virology. 8 (1): 65–73. doi:10.1006/smvy.1997.0107.
10. Daròs JA, Elena SF, Flores R (2006). "Viroids: an Ariadne's thread into the RNA labyrinth". EMBO Rep. 7 (6): 593–8. doi:10.1038/sj.embor.7400706. PMC 1479586 . PMID 16741503.
11. Diener TO (1989). "Circular RNAs: relics of precellular evolution?". Proc. Natl. Acad. Sci. U.S.A. 86 (23): 9370–4. Bibcode:1989PNAS...86.9370D. doi:10.1073/pnas.86.23.9370. PMC 298497 . PMID 2480600.
12. Brian W. J. Mahy, Marc H. V. Van Regenmortel (ed.). Desk Encyclopedia of Plant and Fungal Virology. Academic Press. pp. 71–81. ISBN 978-0123751485.
13. De Bokx JA, Piron PG (1981). "Transmission of potato spindle tuber viroid by aphids.". Netherlands Journal of Plant Pathology. 87 (2): 31–34. doi:10.1007/bf01976653.
14. Flores R, Serra P, Minoia S, Di Serio F, Navarro B (2012). "Viroids: from genotype to phenotype just relying on RNA sequence and structural motifs". Front Microbiol. 3: 217. doi:10.3389/fmicb.2012.00217. PMC 3376415 . PMID 22719735.
15. Hammond RW (1992). "Analysis of the virulence modulating region of potato spindle tuber viroid (PSTVd) by site-directed mutagenesis". Virology. 187 (2): 654–62. doi:10.1016/0042-6822(92)90468-5. PMID 1546460.
16. Wang MB, Bian XY, Wu LM, Liu LX, Smith NA, Isenegger D, Wu RM,

Masuta C, Vance VB, Watson JM, Rezaian A, Dennis ES, Waterhouse PM (2004). "On the role of RNA silencing in the pathogenicity and evolution of viroids and viral satellites". Proc. Natl. Acad. Sci. U.S.A. 101 (9): 3275–80. Bibcode:2004PNAS..101.3275W. doi:10.1073/pnas.0400104101. PMC 365780 . PMID 14978267.

17. Pallas V, Martinez G, Gomez G (2012). "The interaction between plant viroid-induced symptoms and RNA silencing". Methods Mol. Biol. Methods in Molecular Biology. 894: 323–43. doi:10.1007/978-1-61779-882-5_22. ISBN 978-1-61779-881-8. PMID 22678590.
18. Flores R, Gago-Zachert S, Serra P, Sanjuán R, Elena SF (June 18, 2014). "Viroids: survivors from the RNA world?". Annu. Rev. Microbiol. 68: 395–414. doi:10.1146/annurev-micro-091313-103416. PMID 25002087.
19. Zimmer, C (September 25, 2014). "A Tiny Emissary From the Ancient Past". New York Times. Retrieved November 22, 2014.
20. Owens RA, Verhoeven JT (2009). "Potato Spindle Tuber". Plant Health Instructor. doi:10.1094/PHI-I-2009-0804-01.
21. Pommerville, Jeffrey C (2014). Fundamentals of Microbiology. Burlington, MA: Jones and Bartlett Learning. p. 482. ISBN 978-1-284-03968-9.
22. Sänger HL, Klotz G, Riesner D, Gross HJ, Kleinschmidt AK (1976). "Viroids are single-stranded covalently closed circular RNA molecules existing as highly base-paired rod-like structures". Proc. Natl. Acad. Sci. U.S.A. 73 (11): 3852–6. doi:10.1073/pnas.73.11.3852. PMC 431239 . PMID 1069269.
23. Sogo JM, Koller T, Diener TO (1973). "Potato spindle tuber viroid. X. Visualization and size determination by electron microscopy". Virology. 55 (1): 70–80. doi:10.1016/s0042-6822(73)81009-8. PMID 4728831.
24. Gross HJ, Domdey H, Lossow C, Jank P, Raba M, Alberty H, Sänger HL (1978). "Nucleotide sequence and secondary structure of potato spindle tuber viroid". Nature. 273 (5659): 203–8. doi:10.1038/273203a0. PMID 643081.
25. Hammond RW, Owens RA (2006). "Viroids: New and Continuing Risks for Horticultural and Agricultural Crops.". APSnet Feature Articles. doi:10.1094/APSnetFeature-2006-1106.

CHAPTER 7

Phytoplasma

Chapter Contents

Phytoplasma are specialised bacteria that are obligate parasites of plant phloem tissue and transmitting insects (vectors). They were first discovered by scientists in 1967 and were named mycoplasma-like organisms or MLOs. They cannot be cultured in vitro in cell-free media. They are characterised by their lack of a cell wall, a pleiomorphic or filamentous shape, normally with a diameter less than 1 micrometer, and their very small genomes.

INTRODUCTION

Phytoplasmas are specialised bacteria that are obligate parasites of plant phloem tissue and transmitting insects (vectors). They were discovered by scientists in 1967 and were named mycoplasma-like organisms or MLOs. Since their discovery, phytoplasmas have resisted all attempts to be cultured in vitro in any cell free media, hence routine cultivation in artificial media is still to be established. Nevertheless, still under trial stage, phytoplasma growth in specific artificial media has recently been shown. They are characterised by their lack of a cell wall, a pleiomorphic or filamentous shape, normally with a diameter less than 1 μm, and their very small genomes.

Phytoplasmas are pathogens of agriculturally important plants, including coconut, sugarcane, and sandalwood, causing a wide variety of symptoms that range from mild yellowing to death of infected plants. They are most prevalent in tropical and subtropical regions of the world. They require a vector to be transmitted from plant to plant, and this normally takes the form of sap-sucking insects such as leaf hoppers, in which they are also able to survive and replicate.

Phyllody induced by phytoplasma infection on a coneflower (Echinacea purpurea).

References to diseases now known to be caused by phytoplasmas occurred as far back as 1603 for mulberry dwarf disease in Japan. Such diseases were originally thought to be caused by viruses, which, like phytoplasmas, require insect vectors, cannot be cultured, and have some symptom similarity. In 1967, phytoplasmas were discovered in ultrathin sections of plant phloem tissue and named mycoplasma-like organisms (MLOs), because they physically resembled mycoplasmas The organisms were renamed phytoplasmas in 1994, at the 10th Congress of The International Organization for Mycoplasmology.

Morphology

Being Mollicutes, a phytoplasma lacks a cell wall and instead is bound by a triple-layered membrane. The cell membranes of all phytoplasmas studied so far usually contain a single immunodominant protein (of unknown function) that makes up the majority of the protein content of the cell membrane. The typical phytoplasma exhibits a pleiomorphic or filamentous shape and is less than 1 μm in diameter. As prokaryotes, phytoplasmas› DNA is found throughout the cytoplasm, rather than being concentrated in a nucleus.

Symptoms

A common symptom caused by phytoplasma infection is phyllody, the production of leaf-like structures in place of flowers. Evidence suggests the phytoplasma downregulates a gene involved in petal formation (*AP3* and its orthologues) and genes involved in the maintenance of the apical meristem (*Wus* and *CLV1*).Other symptoms, such as the yellowing of leaves, are thought to be caused by the phytoplasma's presence in the phloem, affecting its function and changing the transport of carbohydrates.

Phytoplasma-infected plants may also suffer from virescence, the development of green flowers due to the loss of pigment in the petal cells. Phytoplasma-harboring plants which are able to flower may nevertheless be sterile. A phytoplasma effector protein (SAP54) has been identified as inducing symptoms of virescence and phyllody when expressed in plants.

Many plants infected by phytoplasmas gain a bushy or "witches› broom» appearance due to changes in their normal growth patterns. Most plants show apical dominance, but phytoplasma infection can cause the proliferation of auxiliary (side) shoots and an increase in size of the internodes. Such symptoms are actually useful in the commercial production of poinsettias. The infection produces more axillary shoots, which enables production of poinsettia plants that have more than one flower.

Phytoplasmas may cause many other symptoms that are induced because of the stress placed on the plant by infection rather than specific pathogenicity of the phytoplasma. Photosynthesis, especially photosystem II, is inhibited in many phytoplasma-infected plants. Phytoplasma-infected plants often show yellowing which is caused by the breakdown of chlorophyll, the biosynthesis of which is also inhibited.

Effector (Virulence) Proteins

Many plant pathogens produce virulence factors (or effectors) that modulate or interfere with normal host processes in a way that is beneficial to the pathogen. TCP transcription factors normally regulate plant development and control the expression of lipoxygenase (*LOX*) genes that are required for the biosynthesis of jasmonate. In infected *Arabidopsis* plants (and plants that express SAP11 transgenically), jasmonate levels are decreased. The downregulation of jasmonate production is beneficial to the phytoplasma because jasmonate is involved in plant defence against herbivorous insects such as leafhoppers, and leafhoppers have been shown to lay more eggs on AY-WB-infected plants at least in part because of SAP11. For example, the leafhopper *Macrosteles quadrilineatus* lays 30% more eggs on plants that express SAP11 transgenically, and 60% more eggs on plants infected with AY-WB. Phytoplasmas cannot survive in the external environment and are dependent upon insects such as leafhoppers for transmission to new (healthy) plants. Thus, by interfering with jasmonate production, SAP11 ‹encourages› leafhoppers to lay more eggs on phytoplasma-infected plants, thereby ensuring that newly hatching leafhopper nymphs feed upon infected plants and become vectors for the bacteria.

Transmission

Movement between Plants

Phytoplasmas are mainly spread by insects of the families Cicadellidea (leafhoppers), Fulgoridea (planthoppers), and Psyllidae (jumping plant lice) , which feed on the phloem tissues of infected plants, picking up the phytoplasmas and transmitting them to the next plant on which they feed. So, the host range of phytoplasmas is strongly dependent upon its insect vector. Phytoplasmas contain a major antigenic protein that makes up the majority of their cell surface proteins. This protein has been shown to interact with insect microfilament complexes and is believed to be the determining factor in insect-phytoplasma interaction. Phytoplasmas may overwinter in insect vectors or perennial plants. Phytoplasmas can have varying effects on their insect hosts; examples of both reduced and increased fitness have been seen.

Phytoplasmas enter the insect›s body through the stylet, move through the intestine, and are then absorbed into the haemolymph. From there they proceed to colonise the salivary glands, a process that can take up to three weeks. Once established, phytoplasmas are found in most major organs of an infected insect host. The time between being taken up by the insect and reaching an infectious titre in the salivary glands is called the latency period.

Phytoplasmas can also be spread via dodders (*Cuscuta*) or vegetative propagation such as the grafting of a piece of infected plant onto a healthy plant.

Movement Within Plants

Phytoplasmas are able to move within the phloem from source to sink, and they are able to pass through sieve tube elements. But since they spread more slowly than solutes, for this and other reasons, movement by passive translocation is not supported.

Detection and Diagnosis

Before molecular techniques were developed, the diagnosis of phytoplasma diseases was difficult because they could not be cultured. Thus, classical diagnostic techniques, such as observation of symptoms, were used. Ultrathin sections of the phloem tissue from suspected phytoplasma-infected plants would also be examined for their presence. Treating infected plants with antibiotics such as tetracycline to see if this cured the plant was another diagnostic technique employed.

Molecular diagnostic techniques for the detection of phytoplasma began to emerge in the 1980s and included ELISA-based methods. In the early 1990s, polymerase chain reaction-based methods were developed that were far more sensitive than those that used

ELISA, and RFLP analysis allowed the accurate identification of different strains and species of phytoplasma.

More recently, techniques have been developed that allow for assessment of the level of infection. Both quantitative PCR and bioimaging have been shown to be effective methods of quantifying the titre of phytoplasmas within the plant.

Control

Phytoplasmas are normally controlled by the breeding and planting of disease resistant varieties of crops (believed to the most economically viable option) and by the control of the insect vector.

Tissue culture can be used to produce clones of phytoplasma-infected plants that are healthy. The chances of gaining healthy plants in this manner can be enhanced by the use of cryotherapy, freezing the plant samples in liquid nitrogen, before using them for tissue culture.

Work has also been carried out investigating the effectiveness of plantibodies targeted against phytoplasmas.

Tetracyclines are bacteriostatic to phytoplasmas. However, without continuous use of the antibiotic, disease symptoms reappear. Thus, tetracycline is not a viable control agent in agriculture, but it is used to protect ornamental coconut trees.

Taxonomy

Phytoplasmas belong to the monophyletic order Acholeplasmatales. In 1992, the Subcommittee on the Taxonomy of Mollicutes proposed the use of the name *Phytoplasma* in place of the use of the term MLO (mycoplasma-like organism) "for reference to the phytopathogenic mollicutes». In 2004, the genus name *Phytoplasma* was adopted and is currently at Candidatus status which is used for bacteria that cannot be cultured. Its taxonomy is complicated because it can not be cultured, thus methods normally used for classification of prokaryotes are not possible. Phytoplasma taxonomic groups are based on differences in the fragment sizes produced by the restriction digest of the 16S rRNA gene sequence (RFLP) or by comparison of DNA sequences from the 16s/23s spacer regions. There is some disagreement over how many taxonomic groups the phytoplasmas fall into, recent work involving computer simulated restriction digests of the 16Sr gene suggest there may be up to 28 groups whereas other papers argue for less groups, but more subgroups. Each group includes at least one *Ca.* Phytoplasma species, characterized by distinctive biological, phytopathological, and genetic properties.

A grape vine with "bois noir" phytoplasma disease

A grape vine with "flavescence dorée" phytoplasma disease

Coconut palms dying of lethal yellowing disease

Symptoms of aster yellows on marigold

Tephrosia purpurea witches' broom

Symptoms of elm phloem necrosis phytoplasma

Trees dying of ash yellows phytoplasma

Phyllody caused by phytoplasma infection on Cosmos spp.

Symptoms of sweet potato little leaf phytoplasma on Catharanthus roseus

Phyllody of goldenrod

Sugarcane grassy shoot disease

A palm tree dying of lethal yellowing phytoplasma

A cabbage tree killed by Phytoplasma australiense

MYCOPLASMA

Mycoplasma is a genus of bacteria that lack a cell wall around their cell membrane. Without a cell wall, they are unaffected by many common antibiotics such as penicillin or other beta-lactam antibiotics that target cell wall synthesis. They can be parasitic or saprotrophic. Several species are pathogenic in humans, including M. pneumoniae, which is an important cause of atypical pneumonia and other respiratory disorders, and M. genitalium, which is believed to be involved in pelvic inflammatory diseases. Mycoplasma species are the smallest bacterial cells yet discovered, can survive without oxygen, and come in various shapes. For example, M. genitalium is flask-shaped (about 300 x 600 nm), while M. pneumoniae is more elongated (about 100 x 1000 nm). Hundreds of mycoplasma species infect animals.

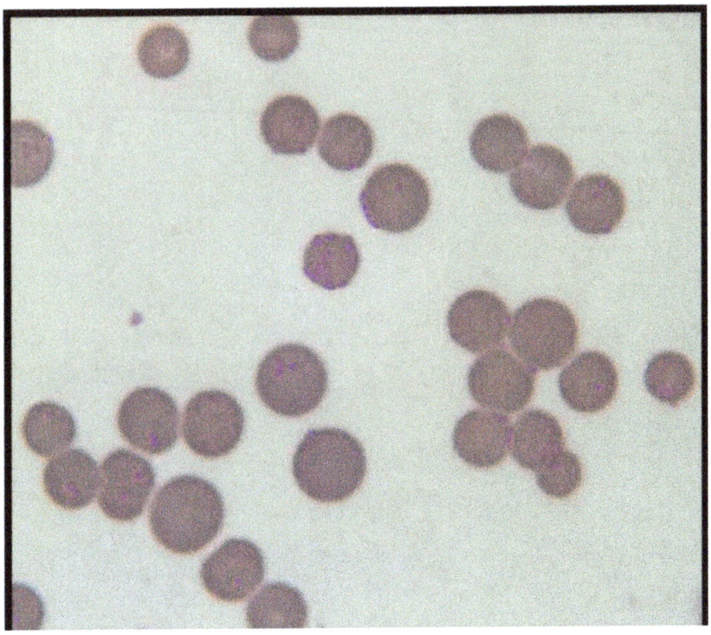

Mycoplasma haemofelis

Origin of the Name

The term *mycoplasma*, from the Greek μυκής, *mykes* (fungus) and πλάσμα, *plasma* (formed), was first used by Albert Bernhard Frank in 1889 to describe an altered state of plant cell cytoplasm resulting from infiltration by fungus-like microorganisms. Julian Nowak later proposed the genus name *Mycoplasma* for certain filamentous microorganisms imagined to have both cellular and acellular stages in their lifecycles, which could explain how they were visible with a microscope, but passed through filters impermeable to bacteria.

Later, the name for *Mycoplasma* was pleuropneumonia-like organisms (PPLO), broadly referring to organisms similar in colonial morphology and filterability to the causative agent (a mycoplasma) of contagious bovine pleuropneumonia.

Species

- M. adleri
- M. agalactiea
- M. agassizii
- M. alkalesens
- M. alligatoris

- M. amphoriforme
- M. anatis
- M. anseris
- M. arginini
- M. arthritidis
- M. auris
- M. bovigenitalium
- M. bovirhinis
- M. bovis
- M. bovoculi
- M. buccale
- M. buteonis
- M. californicum
- M. canadense
- M. canis
- M. capricolum
- M. caviae
- M. cavipharyngis
- M. citelli
- M. cloacale
- M. coccoides
- M. collis
- M. columbinasale
- M. columbinum
- M. columborale
- M. conjunctivae
- M. corogypsi
- M. cottewii
- M. cricetuli
- M. crocodyli
- M. cynos
- M. dispar
- M. edwardii
- M. elephantis
- M. ellychniae

- M. equigenitalium
- M. equirhinis
- M. falconis
- M. fastidiosum
- M. faucium
- M. felifacium
- M. feliminutum
- M. feriruminatoris
- M. fermentans
- M. flocculare
- M. gallinaceum
- M. gallinarum
- M. gallisepticum
- M. gallopavonis
- M. gaeteae
- M. genitalium
- M. glycophilium
- M. gypis
- M. haemocanis
- M. haemofelis
- M. haemomuris
- M. haemosuis
- M. hominis
- M. hyopneumoniae
- M. hypopharyngis
- M. hyorhinis
- M. hyosynoviae
- M. iguanae
- M. imitans
- M. indiense
- M. iners
- M. iowae
- M. laboratorium
- M. lacutcae
- M. lagogenitalium
- M. leachii

- M. leonicptivi
- M. leopharyngis
- M. lipofaciens
- M. lipophilum
- M. lucivorax
- M. luminosum
- M. maculosum
- M. melaleucae
- M. meleagridis
- M. microti
- M. moatsii
- M. mobile
- M. molare
- M. muscosicanis
- M. muris
- M. mustelae
- M. mycoides
- M. neophronis
- M. neurolyticvum
- M. opalescens
- M. orale
- M. ovipneumoniae
- M.ovis
- M. ovipneumoniae
- M. oxoniensis
- M. penetrans
- M. phocae
- M. phocicerebrale
- M. phocidae
- M. phocirhinis
- M. pirum
- M. pneumoniae
- M. primatum
- M. pullorum
- M. pulmonis
- M. putrefaciens

- M. salivarium
- M. simbae
- M. spermatophilum
- M. spumans
- M. sturni
- M. sualvi
- M. subdolum
- M. suis
- M. synoviae
- M. testudineum
- M. testudinis
- M. verecunum
- M. wenyonii
- M. yeatsii

Species that Infect Humans

Other species of Mycoplasma other than those listed below have been recovered from humans, but are assumed to have been contracted from animals. These use humans as the primary host:

- M. amphoriforme
- M. buccale
- M. faucium
- M. fermentans
- M. genitalium
- M. hominis
- M. lipophilum
- M. orale
- M. penetrans
- M. pirum
- M. pneumoniae
- M. primatum
- M. salivarium
- M. spermatophilum

Pathophysiology

Mycoplasma species have been isolated from women with bacterial vaginosis. *M. genitalium* infection is associated with increased risk of cervicitis, pelvic inflammatory disease, preterm birth and spontaneous abortion, and infertility. Mycoplasmae are associated with fetal respiratory distress syndrome, bronchopulmonary dysplasia, and intraventricular hemorrhage in preterm infants.

Characteristics

Over 100 species have been included in the genus *Mycoplasma*. Microbes of the class Mollicutes, to which Mycoplasma belongs, are parasites or commensals of humans, animals, and plants. The genus *Mycoplasma* uses vertebrate and arthropod hosts. Dietary nitrogen availability has been shown to alter codon bias and genome evolution in *Mycoplasma* and *Phytoplasma*.

Cell morphology

"Due to the lack of a rigid cell wall, Mycoplasmataceae can contort into a broad range of shapes, from round to oblong. They therefore cannot be classified as rods or cocci."

Taxonomy

Laboratory Contaminant

Mycoplasma species are often found in research laboratories as contaminants in cell culture. Mycoplasmal cell culture contamination occurs due to contamination from individuals or contaminated cell culture medium ingredients. *Mycoplasma* cells are physically small – less than 1 µm, so are difficult to detect with a conventional microscope.

Mycoplasmae may induce cellular changes, including chromosome aberrations, changes in metabolism and cell growth. Severe *Mycoplasma* infections may destroy a cell line. Detection techniques include DNA Probe, enzyme immunoassays, PCR, plating on sensitive agar and staining with a DNA stain including DAPI or Hoechst.

An estimated 11 to 15% of U.S. laboratory cell cultures are contaminated with mycoplasma. A Corning study showed that half of U.S. scientists did not test for *Mycoplasma* contamination in their cell cultures. The study also stated that, in former Czechoslovakia, 100% of cell cultures that were not routinely tested were contaminated while only 2% of those routinely tested were contaminated (study page 6). Since the U.S. contamination rate was based on a study of companies that routinely checked for *Mycoplasma*, the actual contamination rate may be higher. European contamination rates are higher and that of other countries are higher still (up to 80% of Japanese cell cultures). About 1% of published Gene Expression Omnibus data may have been compromised. Several antibiotic-containing formulations of antimycoplasmal reagents have been developed over the years.

Synthetic Mycoplasma Genome

A chemically synthesized genome of a mycoplasmal cell based entirely on synthetic DNA which can self-replicate has been referred to as *Mycoplasma laboratorium*.

Pathogenicity

The P1 antigen is the primary virulence factor of mycoplasma . P1 is a membrane associated protein that allows adhesion to epithelial cells. The P1 receptor is also expressed on erythrocytes which can lead to autoantibody agglutination from mycobacteria infection. Several *Mycoplasma* species can cause disease, including *M. pneumoniae*, which is an important cause of atypical pneumonia (formerly known as "walking pneumonia»), and *M. genitalium*, which has been associated with pelvic inflammatory diseases. Mycoplasma infections in humans are associated with skin eruptions in 17% of cases.

Mycoplasma Infection and Host Cell Transformation

The presence of *Mycoplasma* was first reported in samples of cancer tissue in the 1960s. Since then, several studies tried to find and prove the connection between *Mycoplasma* and cancer, as well as how the bacterium might be involved in the formation of cancer. Several studies have shown that cells that are chronically infected with the bacteria go through a multistep transformation. The changes caused by chronic mycoplasmal infections occur gradually and are both morphological and genetic. The first visual sign of infection is when the cells gradually shift from their normal form to sickle-shaped. They also become hyperchromatic due to an increase of DNA in the nucleus of the cells. In later stages, the cells lose the need for a solid support to grow and proliferate, as well as the normal contact-dependent inhibition cells.

Possible Intracellular Mechanisms of Mycoplasmal Malignant Transformation

Karyotypic Changes Related to Mycoplasma Infections

Cells infected with *Mycoplasma* for an extended period of time show significant chromosomal abnormalities. These include the addition of chromosomes, the loss of entire chromosomes, partial loss of chromosomes, and chromosomal translocation. All of these genetic abnormalities may contribute to the process of malignant transformation. Chromosomal translocation and extra chromosomes help create abnormally high activity of certain proto-oncogenes, which caused by these genetic abnormalities and include those encoding c-myc, HRAS, and vav. The activity of proto-oncogenes is not the only cellular function that is affected; tumour suppressor genes are affected by the chromosomal changes induced by mycoplasma, as well. Partial or complete loss of chromosomes causes the loss of important genes involved in the regulation of cell proliferation. Two genes whose activities are markedly decreased during chronic infections with mycoplasma are the Rb and the p53 tumour suppressor genes. Another possible mechanism of carcinogenesis is RAC1 activation by a small GTPase-like protein fragment of Mycoplasma. A major feature that differentiates mycoplasmas from other carcinogenic pathogens is that the mycoplasmas do not cause the cellular changes by insertion of their own genetic material into the host cell. The exact mechanism by which the bacterium causes the changes is not yet known.

Partial Reversibility of Malignant Transformations

The malignant transformation induced by mycoplasmae is also different from that caused by other pathogens in that the process is reversible. The state of reversal is, however, only possible up to a certain point during the infection. The window of time when reversibility is possible varies greatly; it depends primarily on the *Mycoplasma* involved. In the case of *M. fermentans*, the transformation is reversible until around week 11 of infection and starts to become irreversible between weeks 11 and 18. If the bacteria are killed using antibiotics (i.e. ciprofloxacin or Clarithromycin) before the irreversible stage, the infected cells should return to normal.

REFERENCES

1. Ryan KJ, Ray CG (editors) (2004). Sherris Medical Microbiology (4th ed.). McGraw Hill. pp. 409–12. ISBN 0-8385-8529-9.
2. Richard L. Sweet; Ronald S. Gibbs. Infectious Diseases of the Female Genital Tract. Lippincott Williams & Wilkins, 2009.
3. Larsen, Bryan; Hwang, Joseph (2010). "Mycoplasma, Ureaplasma, and Adverse Pregnancy Outcomes: A Fresh Look". Infectious Diseases in Obstetrics and Gynecology. 2010: 1–7. doi:10.1155/2010/521921. ISSN 1064-7449.
4. Krass CJ, Gardner MW (January 1973). "Etymology of the Term Mycoplasma" (PDF). Int. J. Of Syst. Bact. 23 (1): 62–64. doi:10.1099/00207713-23-1-62.
5. Browning GF, Citti C (editors) (2014). Mollicutes Molecular Biology and Pathogenesis (1st ed.). Caister Academic Press. pp. 1–14. ISBN 978-1-908230-30-0.
6. Edward DG, Freundt EA (February 1956). "The classification and nomenclature of organisms of the pleuropneumonia group"(PDF). J. Gen. Microbiol. 14 (1): 197–207. doi:10.1099/00221287-14-1-197. PMID 13306904.
7. LPSN bacterio.net
8. Waites, K. B.; Katz, B.; Schelonka, R. L. (2005). "Mycoplasmas and Ureaplasmas as Neonatal Pathogens". Clinical Microbiology Reviews. 18 (4): 757–789. doi:10.1128/CMR.18.4.757-789.2005. ISSN 0893-8512.
9. Lis, R.; Rowhani-Rahbar, A.; Manhart, L. E. (2015). "Mycoplasma genitalium Infection and Female Reproductive Tract Disease: A Meta-Analysis". Clinical Infectious Diseases. 61 (3): 418–426. doi:10.1093/cid/civ312. ISSN 1058-4838. PMID 25900174.
10. Maggi, Ricardo G.; Compton, Sarah M.; Trull, Chelsea L.; Mascarelli, Patricia

E.; Mozayeni, B. Robert; Breitschwerdt, Edward B. (1 October 2013). "Infection with Hemotropic Mycoplasma Species in Patients with or without Extensive Arthropod or Animal Contact". Journal of Clinical Microbiology. pp. 3237–3241. doi:10.1128/JCM.01125-13.

11. Seward, Emily; Kelly, Steve (2016). "Dietary nitrogen alters codon bias and genome composition in parasitic microorganisms". Genome Biology. 17 (226): 3–15. doi:10.1186/s13059-016-1087-9. PMC 5109750 . PMID 27842572.
12. Gladwin, MD, Mark; Trattler, MD, William; Mahan, MD, C. Scott (2014). Clinical Microbiology made ridiculously simple. Miami, Fl: MedMaster, Inc. p. 156. ISBN 978-1935660156.
13. Drexler HG; Uphoff CC (2002). "Mycoplasma contamination of cell cultures: Incidence, sources, effects, detection, elimination, prevention". Cytotechnology. 39 (2): 75–90. doi:10.1023/A:1022913015916. PMC 3463982 . PMID 19003295.
14. John Ryan (2008). "Understanding and Managing Cell Culture Contamination" (PDF). Corning Incorporated. p. 24.
15. Aldecoa-Otalora E, Langdon W, Cunningham P, Arno MJ (December 2009). "Unexpected presence of mycoplasma probes on human microarrays". BioTechniques. 47 (6): 1013–5. doi:10.2144/000113271. PMID 20047202.
16. Link into RNAnet showing contamination of GEO. Press plot and drag blue crosshairs to expose links to description of experiments on human RNA samples
17. BM-Cyclin by Roche, MRA by ICN, Plasmocin by Invivogen and more recently De-Plasma by TOKU-E.
18. Gibson DG, Glass JI, Lartigue C, Noskov VN, Chuang RY, Algire MA, Benders GA, Montague MG, Ma L, Moodie MM, Merryman C, Vashee S, Krishnakumar R, Assad-Garcia N, Andrews-Pfannkoch C, Denisova EA, Young L, Qi ZQ, Segall-Shapiro TH, Calvey CH, Parmar PP, Hutchison CA, Smith HO, Venter JC (July 2010). "Creation of a bacterial cell controlled by a chemically synthesized genome". Science. 329 (5987): 52–6. doi:10.1126/science.1190719. PMID 20488990.
19. Parija, Subhash Chandra (2014). Textbook of Microbiology & Immunology. Elsevier Health Sciences. ISBN 9788131236246.
20. James, William D.; Berger, Timothy G.; et al. (2006). Andrews' Diseases of the Skin: clinical Dermatology. Saunders Elsevier. ISBN 0-7216-2921-0.
21. Ljubin-Sternak, Suncanica; Mestrovic, Tomislav (2014). "Review: Clamydia trachonmatis and Genital Mycoplasmias: Pathogens with an Impact on Human Reproductive Health". Journal of Pathogens. 2014 (183167): 1–15. doi:10.1155/2014/183167. PMC 4295611 . PMID 25614838.
22. Huang S, Li JY, Wu J, Meng L, Shou CC (April 2001). "Mycoplasma infections and different human carcinomas". World Journal of Gastroenterology. 7 (2): 266–269. doi:10.3748/wjg.v7.i2.266. PMID 11819772.
23. Sinkovics JG (February 2012). "Molecular biology of oncogenic inflammatory processes. I. Non-oncogenic and oncogenic pathogens, intrinsic inflammatory

reactions without pathogens, and microRNA/DNA interactions (Review)". International Journal of Oncology. 40 (2): 305–349. doi:10.3892/ijo.2011.1248. PMID 22076306.

24. Tsai S, Wear DJ, Shih JW, Lo SC (October 1995). "Mycoplasmas and oncogenesis: Persistent infection and multistage malignant transformation". Proceedings of the National Academy of Sciences of the United States of America. 92 (22): 10197–10201. doi:10.1073/pnas.92.22.10197. PMID 7479753.
25. Cimolai N (August 2001). "Do mycoplasmas cause human cancer?". Canadian Journal of Microbiology. 47 (8): 691–697. doi:10.1139/w01-053. PMID 11575494.
26. Jiang S, Zhang S, Langenfeld J, Lo SC, Rogers MB (May 2008). "Mycoplasma infection transforms normal lung cells and induces bone morphogenetic protein 2 expression by post-transcriptional mechanisms". Journal of Cellular Biochemistry. 104 (2): 580–594. doi:10.1002/jcb.21647. PMID 18059017.
27. Zhang S, Tsai S, Lo SC (May 2006). "Alteration of gene expression profiles during mycoplasma-induced malignant cell transformation". BMC Cancer. 6: 116. doi:10.1186/1471-2407-6-116. PMID 16674811.
28. Namiki K, Goodison S, Porvasnik S, Allan RW, Iczkowski KA, Urbanek C, Reyes L, Sakamoto N, Rosser CJ (September 2009). "Persistent exposure to mycoplasma induces malignant transformation of human prostate cells". PLoS ONE. 4 (9): 1–9. doi:10.1371/journal.pone.0006872. PMC 2730529 . PMID 19721714.
29. Barykova YA, Logunov DY, Shmarov MM, Vinarov AZ, Fiev DN, Vinarova NA, Rakovskaya IV, Baker PS, Shyshynova I, Stephenson AJ, Klein EA, Naroditsky BS, Gintsburg AL, Gudkov AV (2011). "Association of Mycoplasma hominis infection with prostate cancer". Oncotarget. 2: 289–97. doi:10.18632/oncotarget.256. PMC 3248169 . PMID 21471611.
30. Chan PJ, Seraj IM, Kalugdan TH, King A (November 1996). "Prevalence of mycoplasma conserved DNA in malignant ovarian cancer detected using sensitive PCR–ELISA". Gynecologic Oncology. 63 (2): 258–260. doi:10.1006/gyno.1996.0316. PMID 8910637.
31. Xiaolei C, Taot H, Zongli S, Hongying Y (2014). "The role of ureaplasma urealyticum infection in cervical intraepithelial neoplasia and cervical cancer". European Journal of Gynaecological Oncology. 35 (5): 571–5. PMID 25423707.
32. Hu X, Yu J, Zhou X, Li Z, Xia Y, Luo Z, Wu Y (Jan 2014). "A small GTPase-like protein fragment of Mycoplasma promotes tumor cell migration and proliferation in vitro via interaction with Rac1 and Stat3.". J Mol Med Rep. 9 (1): 173–179. doi:10.3892/mmr.2013.1766. PMID 24172987.
33. Pehlivan M, Pehlivan S, Onay H, Koyuncuoglu M, Kirkali Z (February 2005). "Can mycoplasma-mediated oncogenesis be responsible for formation of conventional renal cell carcinoma?". Urology. 65 (2): 411–414. doi:10.1016/j.urology.2004.10.015. PMID 15708077.
34. Mariotti E, Gemei M, Mirabelli P, D'Alessio F, Di Noto R, Fortunato G, Del Vecchio L (March 2010). "The percentage of CD133+ cells in human colorectal

cancer cell lines is influenced by Mycoplasma hyorhinis infection". BMC Cancer. 10: 120–125. doi:10.1186/1471-2407-10-120. PMID 20353562.

35. Yang H, Qu L, Ma H, Chen L, Liu W, Liu C, Meng L, Wu J, Shou C (November 2010). "Mycoplasma hyorhinis infection in gastric carcinoma and its effects on the malignant phenotypes of gastric cancer cells". BMC Gastroenterology. 10: 132–140. doi:10.1186/1471-230X-10-132. PMID 21062494.
36. Apostolou P, Tsantsaridou A, Papasotiriou I, Toloudi M, Chatziioannou M, Giamouzis G (October 2011). "Bacterial and fungal microflora in surgically removed lung cancer samples". Journal of Cardiothoracic Surgery. 6: 137. doi:10.1186/1749-8090-6-137. PMID 21999143.
37. Urbanek C, Goodison S, Chang M, Porvasnik S, Sakamoto N, Li CZ, Boehlein SK, Rosser CJ (June 2011). "Detection of antibodies directed at M. hyorhinis p37 in the serum of men with newly diagnosed prostate cancer". BMC Cancer. 11 (1): 233–238. doi:10.1186/1471-2407-11-233. PMID 21663671.
38. Doi Y, Teranaka M, Yora K, Asuyama H (1967). "Mycoplasma or PLT-group-like organisms found in the phloem elements of plants infected with mulberry dwarf, potato witches' broom, aster yellows or paulownia witches' broom". Annals of the Phytopathological Society of Japan. 33 (4): 259–266. doi:10.3186/jjphytopath.33.259.
39. Contaldo, N.; Bertaccini, A.; Paltrinieri, S.; Windsor, H.M.; Windsor, G.D. (2012). "Axenic culture of a plant pathogenic Phytoplasma". Phytopathologia Mediterranea. 244: 607–117. doi:10.14601/Phytopathol_Mediterr-11773.
40. Okuda, S (1972). "Occurrence of diseases caused by mycoplasma-like organisms in Japan". Plant Protection. 26: 180–183.
41. Hogenhout, SA; Oshima K; Ammar E-D; Kakizawa S; Kingdom HN; Namba S (2008). "Phytoplasmas: bacteria that manipulate plants and insects". Molecular Plant Pathology. 9 (4): 403–423. doi:10.1111/j.1364-3703.2008.00472.x. PMID 18705857. Retrieved 2008-07-04.
42. Bertamini, M; Grando M. S; Nedunchezhian N (2004). "Effects of Phytoplasma Infection on Pigments, Chlorophyll-Protein Complex and Photosynthetic Activities in Field Grown Apple Leaves". Biologia Plantarum. Springer Netherlands. 47 (2): 237–242. doi:10.1006/pmpp.2003.0450.
43. Berg, M; Davis DL; Clark MF; Vetten HJ; Maier G; Marcone C; Seemuller E (1999). "Isolation of the gene encoding an immunodominant membrane protein of the apple proliferation phytoplasma, and expression and characterization of the gene product". Microbiology. Society for General Microbiology. 145: 1939–1943. doi:10.1099/13500872-145-8-1937. PMID 10463160.
44. Pracros, P; Renaudin J; Eveillard S; Mouras A; Hernould M (2006). "Tomato Flower Abnormalities Induced by Stolbur Phytoplasma Infection Are Associated with Changes of Expression of Floral Development Genes". Molecular Plant Microbe Interactions. APS Press. 19 (1): 62–68. doi:10.1094/MPMI-19-0062. PMID 16404954.
45. Muast BE, Espadas F, Talavera C, Aguilar M, Santamaría JM, Oropeza C

(2003). "Changes in carbohydrate metabolism in coconut palms infected with the lethal yellowing phytoplasma". Phytopathology. 93 (8): 976–981. doi:10.1094/PHYTO.2003.93.8.976. PMID 18943864.

46. Lee, IM; Davis RE; Gundersen-Rindal DE (2000). "Phytoplasma: Phytopathogenic Mollicutes". Annual Review of Microbiology. 54: 221–255. doi:10.1146/annurev.micro.54.1.221. PMID 11018129.
47. Lee, IM; Klopmeyer M; Bartoszyk IM; Gundersen-Rindal DE; Chou TS; Thomson KL; Eisenreich R (1997). "Phytoplasma induced free-branching in commercial poinsettia cultivars". Nature Biotechnology. Nature Publishing Group. 15 (2): 178–182. doi:10.1038/nbt0297-178. PMID 9035146.
48. Sugio, Akiko; Kingdom, H. N.; MacLean, A. M.; Grieve, V. M.; Hogenhout, S. A. (2011). "Phytoplasma protein effector SAP11 enhances insect vector reproduction by manipulating plant development and defense hormone biosynthesis". Proceedings of the National Academy of Sciences. 108 (48): E1254–E1263. doi:10.1073/pnas.1105664108.
49. Weintraub, Phyllis G.; Beanland, LeAnn (2006). "Insect vectors of phytoplasmas". Annual Review of Entomology. 51: 91–111. doi:10.1146/annurev.ento.51.110104.151039. PMID 16332205.
50. Suzuki, S.; Oshima, K.; Kakizawa, S.; Arashida, R.; Jung, H.-Y.; Yamaji, Y.; Nishigawa, H.; Ugaki, M.; Namba, S. (2006). "Interactions between a membrane protein of a pathogen and insect microfilament complex determines insect vector specificity". Proceedings of the National Academy of Sciences. 103 (11): 4252–4257. doi:10.1073/pnas.0508668103.
51. Christensen N, Axelsen K, Nicolaisen M, Schulz A (2005). "Phytoplasmas and their interactions with their hosts". Trends in Plant Sciences. 10 (11): 526–535. doi:10.1016/j.tplants.2005.09.008.
52. Carraro, L.; Loi, N.; Favali, M. A.; Favali, M. A. (1991). "Transmission characteristics of the clover phyllody agent by dodder". J. Phytopathol. 133: 15–22. doi:10.1111/j.1439-0434.1991.tb00132.x.
53. Christensen NM, Nicolaisen M, Hansen M, Schulz A (2004). "Distribution of phytoplasmas in infected plants as revealed by real time PCR and bioimaging". Molecular Plant Microbe Interactions. 17 (11): 1175–1184. doi:10.1094/MPMI.2004.17.11.1175. PMID 15553243.
54. Chen; et al. (1992). "Detection and identification of plant and insect mollicutes". The Mycoplasmas. 5: 393–424.
55. Wang et al. (2007) Effective elimination of sweet potato little lead by cryotherapy of shoot tips. Plant Pathology online early edition.
56. Chen, Y. D.; Chen, T. A. (1998). "Expression of engineered antibodies in plants: A possible tool for spiroplasma and phytoplasma disease control". Phytopathology. 88 (12): 1367–1371. doi:10.1094/PHYTO.1998.88.12.1367. PMID 18944841.
57. Davies, R. E.; Whitcomb, R. F.; Steere, R. L. (1968). "Remission of aster yellows disease by antibiotics". Science. 161 (3843): 793–794. doi:10.1126/science.161.3843.793.

58. Drug for Humans Checks Palm Trees Disease. New York Times, July 19 1983

59. Bai X, Zhang J, Ewing A, Miller SA, Jancso Radek A, Shevchenko DV, Tsukerman K, Walunas T, et al. (2006). "Living with Genome Instability: the Adaptation of Phytoplasmas to Diverse Environments of Their Insect and Plant Hosts". Journal of Bacteriology. 188 (10): 3682–3696. doi:10.1128/JB.188.10.3682-3696.2006. PMC 1482866 . PMID 16672622.

60. Oshima K, Kakizawa S, Nishigawa H, Jung HY, Wei W, Suzuki S, Arashida R, Nakata D, et al. (2004). "Reductive evolution suggested from the complete genome sequence of a plant-pathogenic phytoplasma". Nature Genetics. 36 (1): 27–29. doi:10.1038/ng1277. PMID 14661021.

61. Tran-Nguyen, LT; Kube, M; Schneider, B; Reinhardt, R; Gibb, KS (2008). "Comparative Genome Analysis of "Candidatus Phytoplasma australiense" (Subgroup tuf-Australia I; rp-A) and "Ca. Phytoplasma asteris" Strains OY-M and AY-WB". Journal of Bacteriology. 190 (11): 3979–91. doi:10.1128/JB.01301-07. PMC 2395047 . PMID 18359806.

62. Dikinson, M. Molecular Plant Pathology (2003) BIOS Scientific Publishers

63. Marcone, C; Neimark H; Ragozzino A; Lauer U; Seemüller E (1999). "Chromosome Sizes of Phytoplasmas Composing Major Phylogenetic Groups and Subgroups". Phytopathology. APS Press. 89 (9): 805–810. doi:10.1094/PHYTO.1999.89.9.805. PMID 18944709.

64. Nishigawa; et al. (2003). "Complete set of extrachromosomal DNAs from three pathogenic lines of onion yellows phytoplasma and use of PCR to differentiate each line". Journal of General Plant Pathology. 69: 194–198.

65. Razin S, Yogev D, Naot Y (1998). "Molecular Biology and Pathogenicity of Mycoplasmas". Microbiology Molecular Biology Review. 62 (4): 1094–1156. PMC 98941 . PMID 9841667.

66. Jomantiene R, Davis RE (2006). "Clusters of diverse genes existing as multiple, sequence variable mosaics in a phytoplasma genomes". FEMS Microbiology Letters. 255 (1): 59–65. doi:10.1111/j.1574-6968.2005.00057.x. PMID 16436062.

67. Subcommittee on the Taxonomy of Mollicutes. Minutes of the Interim Meetings, 1 and 2 August, 1992, Ames, IowaInt. J. of Syst. Bact. April 1993, p. 394-397; Vol. 43, No. 2 (see minutes 10 and 25)

68. The IRPCM Phytoplasma/Spiroplasma Working Team - Phytoplasma taxonomy group: Candidatus Phytoplasma, a taxon for the wall-less, non-helical prokaryotes that colonize plant phloem and insects. Int. J. Syst. Evol. Microbiol., 2004, 54, 1243-1255.

69. Murry; Syst, Bacteriol; et al. (1995). 45: 186–187. Missing or empty |title= (help)[full citation needed]

70. Hodgetts, J.; Ball, T.; Boonham, N.; Mumford, R.; Dickinson, M. (2007). "Taxonomic groupings based on the analysis on the 16s/23s spacer regions which shows greater variation than the normally used 16srRNA gene results in classification similar to that derived from 16s rRNA data but with more detailed subdivisions". Plant Pathology. 56: 357–365. doi:10.1111/j.1365-

3059.2006.01561.x.

71. Wei W, Davis RE, Lee IM, Zhao Y (2007). "Computer-simulated RFLP analysis of 16S rRNA genes: identification of ten new phytoplasma groups". International Journal of Systematic and Evolutionary Microbiology. 57 (Pt 8): 1855–1867. doi:10.1099/ijs.0.65000-0. PMID 17684271.

72. Yadav, A.; Bhale, U.; Thorat, V.; Shouche, Y. (2014). "First Report of new subgroup 16SrII- M 'Candidatus Phytoplasma aurantifolia' associated with 'Witches Broom' disease of Tephrosia purpurea in India". Plant Disease. 98: 990. doi:10.1094/PDIS-11-13-1183-PDN.

73. Nasare, K.; Yadav, Amit; Singh, A. K.; Shivasharanappa, K. B.; Nerkar, Y. S.; Reddy, V. S. (2007). "Molecular and symptom analysis reveal the presence of new phytoplasmas associated with sugarcane grassy shoot disease in India". Plant Disease. 91 (11): 1413–1418. doi:10.1094/PDIS-91-11-1413.

CHAPTER 8

Fungus

Chapter Contents

- **Introduction**
- **Etymology**
- **Mycology**
- **Morphology**
- **Growth and Physiology**
- **Reproduction**
- **Taxonomy**

A fungus is neither a plant nor an animal. It is similar to a plant, but it has no chlorophyll and cannot make its own food like a plant can through photosynthesis. They get their food by absorbing nutrients from their surroundings.

INTRODUCTION

A fungus is any member of the group of eukaryotic organisms that includes unicellular microorganisms such as yeasts and molds, as well as multicellular fungi that produce familiar fruiting forms known as mushrooms. These organisms are classified as a kingdom, Fungi, which is separate from the other eukaryotic life kingdoms of plants and animals.

A characteristic that places fungi in a different kingdom from plants, bacteria and some protists, is chitin in their cell walls. Similar to animals, fungi are heterotrophs; they acquire their food by absorbing dissolved molecules, typically by secreting digestive enzymes into their environment. Fungi do not photosynthesise. Growth is their means of mobility, except for spores (a few of which are flagellated), which may travel through the air or water. Fungi are the principal decomposers in ecological systems. These and other differences place fungi in a single group of related organisms, named the Eumycota (true fungi or Eumycetes), which share a common ancestor (is a monophyletic group), an interpretation that is also strongly supported by molecular phylogenetics. This fungal group is distinct from the structurally similar myxomycetes (slime molds) and oomycetes (water molds). The discipline of biology devoted to the study of fungi is known as mycology (from the Greek μύκης, mukēs, meaning "fungus"). In the past, mycology was regarded as a branch of botany, although it is now known fungi are genetically more closely related to animals than to plants.

Abundant worldwide, most fungi are inconspicuous because of the small size of their structures, and their cryptic lifestyles in soil or on dead matter. Fungi include symbionts of plants, animals, or other fungi and also parasites. They may become noticeable when fruiting, either as mushrooms or as molds. Fungi perform an essential role in the decomposition of organic matter and have fundamental roles in nutrient cycling and exchange in the environment. They have long been used as a direct source of food, in the form of mushrooms and truffles; as a leavening agent for bread; and in the fermentation of various food products, such as wine, beer, and soy sauce. Since the 1940s, fungi have been used for the production of antibiotics, and, more recently, various enzymes produced by fungi are used industrially and in detergents. Fungi are also used as biological pesticides to control weeds, plant diseases and insect pests. Many species produce bioactive compounds called mycotoxins, such as alkaloids and polyketides, that are toxic to animals including humans. The fruiting structures of a few species contain psychotropic compounds and are consumed recreationally or in traditional spiritual ceremonies. Fungi can break down manufactured materials and buildings, and become significant pathogens of humans and other animals. Losses of crops due to fungal diseases (e.g., rice blast disease) or food spoilage can have a large impact on human food supplies and local economies.

The fungus kingdom encompasses an enormous diversity of taxa with varied ecologies, life cycle strategies, and morphologies ranging from unicellular aquatic chytrids to large mushrooms. However, little is known of the true biodiversity of Kingdom Fungi, which has been estimated at 1.5 million to 5 million species, with about 5% of these having been formally classified. Ever since the pioneering 18th and 19th century taxonomical works of Carl Linnaeus, Christian Hendrik Persoon, and Elias Magnus Fries, fungi have been classified according to their morphology (e.g., characteristics such as spore color or microscopic features) or physiology. Advances in molecular genetics have opened the way for DNA analysis to be incorporated into taxonomy, which has sometimes challenged the historical groupings based on morphology and other traits. Phylogenetic studies published in the last decade have helped reshape the classification within Kingdom Fungi, which is divided into one subkingdom, seven phyla, and ten subphyla.

ETYMOLOGY

The English word *fungus* is directly adopted from the Latin *fungus* (mushroom), used in the writings of Horace and Pliny.This in turn is derived from the Greek word *sphongos* (σφογγος "sponge"), which refers to the macroscopic structures and morphology of mushrooms and molds; the root is also used in other languages, such as the German *Schwamm* ("sponge") and *Schimmel* ("mold"). The use of the word *mycology*, which is derived from the Greek *mykes* (μύκης "mushroom") and *logos* (λόγος "discourse"), to denote the scientific study of fungi is thought to have originated in 1836 with English naturalist Miles Joseph Berkeley's publication *The English Flora of Sir James Edward Smith, Vol. 5.* A group of all the fungi present in a particular area or geographic region is known as mycobiota (plural noun, no singular), e.g., "the mycobiota of Ireland".

Characteristics

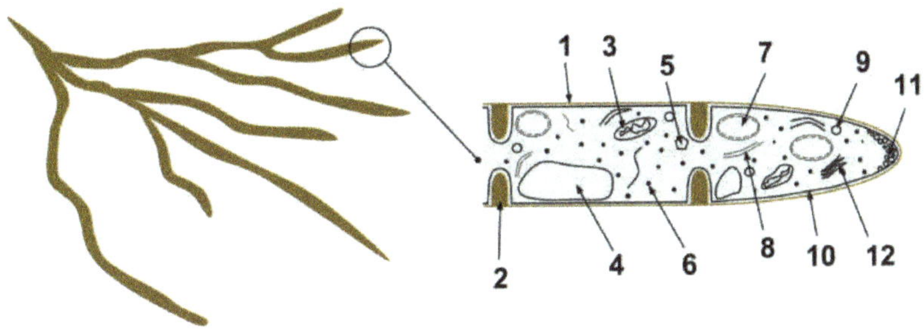

Fungal Hyphae Cells

- Hyphal wall
- Septum
- Mitochondrion
- Vacuole
- Ergosterol crystal
- Ribosome
- Nucleus
- Endoplasmic reticulum
- Lipid body
- Plasma membrane
- Spitzenkörper
- Golgi apparatus

Before the introduction of molecular methods for phylogenetic analysis, taxonomists considered fungi to be members of the plant kingdom because of similarities in lifestyle: both fungi and plants are mainly immobile, and have similarities in general morphology and growth habitat. Like plants, fungi often grow in soil and, in the case of mushrooms, form conspicuous fruit bodies, which sometimes resemble plants, such as mosses. The fungi are now considered a separate kingdom, distinct from both plants and animals, from which they appear to have diverged around one billion years ago. Some morphological, biochemical, and genetic features are shared with other organisms, while others are unique to the fungi, clearly separating them from the other kingdoms:

Shared features:

- With other eukaryotes: Fungal cells contain membrane-bound nuclei with chromosomes that contain DNA with noncoding regions called introns and coding regions called exons. Fungi have membrane-bound cytoplasmic organelles such as mitochondria, sterol-containing membranes, and ribosomes of the 80S type. They have a characteristic range of soluble carbohydrates and storage compounds, including sugar alcohols (e.g., mannitol), disaccharides, (e.g., trehalose), and polysaccharides (e.g., glycogen, which is also found in animals).
- With animals: Fungi lack chloroplasts and are heterotrophic organisms and so require preformed organic compounds as energy sources.
- With plants: Fungi have a cell wall and vacuoles. They reproduce by both sexual and asexual means, and like basal plant groups (such as ferns and mosses) produce spores. Similar to mosses and algae, fungi typically have haploid nuclei.
- With euglenoids and bacteria: Higher fungi, euglenoids, and some bacteria produce the amino acid L-lysine in specific biosynthesis steps, called the α-aminoadipate pathway.
- The cells of most fungi grow as tubular, elongated, and thread-like (filamentous) structures called hyphae, which may contain multiple nuclei and extend by growing at their tips. Each tip contains a set of aggregated vesicles—cellular structures consisting of proteins, lipids, and other organic molecules—called the Spitzenkörper. Both fungi and oomycetes grow as filamentous hyphal cells. In contrast, similar-looking organisms, such as filamentous green algae, grow by repeated cell division within a chain of cells. There are also single-celled fungi (yeasts) that do not form hyphae, and fungi with both hyphal and yeast forms.
- In common with some plant and animal species, more than 70 fungal species display bioluminescence.

Unique Features

- Some species grow as unicellular yeasts that reproduce by budding or binary fission. Dimorphic fungi can switch between a yeast phase and a hyphal phase in response to environmental conditions.
- The fungal cell wall is composed of glucans and chitin; while glucans are also found in plants and chitin in the exoskeleton of arthropods, fungi are the only organisms that combine these two structural molecules in their cell wall. Unlike those of plants and oomycetes, fungal cell walls do not contain cellulose.

Most fungi lack an efficient system for the long-distance transport of water and nutrients, such as the xylem and phloem in many plants. To overcome this limitation, some fungi, such as *Armillaria*, form rhizomorphs, which resemble and perform functions similar to the roots of plants. As eukaryotes, fungi possess a biosynthetic pathway for producing terpenes that uses mevalonic acid and pyrophosphate as chemical building blocks. Plants and some other organisms have an additional terpene biosynthesis pathway in their chloroplasts, a structure fungi and animals do not have. Fungi produce several secondary metabolites that are similar or identical in structure to those made by plants. Many of the plant and fungal enzymes that make these compounds differ from each other in sequence and other characteristics, which indicates separate origins and evolution of these enzymes in the fungi and plants.

Omphalotus nidiformis, a bioluminescent mushroom

Diversity

Bracket fungi on a tree stump

Fungi have a worldwide distribution, and grow in a wide range of habitats, including extreme environments such as deserts or areas with high salt concentrations or ionizing radiation, as well as in deep sea sediments. Some can survive the intense UV and cosmic radiation encountered during space travel. Most grow in terrestrial environments, though several species live partly or solely in aquatic habitats, such as the chytrid fungus *Batrachochytrium dendrobatidis*, a parasite that has been responsible for a worldwide decline in amphibian populations. This organism spends part of its life cycle as a motile zoospore, enabling it to propel itself through water and enter its amphibian host. Other examples of aquatic fungi include those living in hydrothermal areas of the ocean.

Around 100,000 species of fungi have been formally described by taxonomists, but the global biodiversity of the fungus kingdom is not fully understood. On the basis of observations of the ratio of the number of fungal species to the number of plant species in selected environments, the fungal kingdom has been estimated to contain about 1.5 million species. A recent (2011) estimate suggests there may be over 5 million species. In mycology, species have historically been distinguished by a variety of methods and concepts. Classification based on morphological characteristics, such as the size and shape of spores or fruiting structures, has traditionally dominated fungal taxonomy. Species may also be distinguished by their biochemical and physiological characteristics, such as their ability to metabolize certain biochemicals, or their reaction to chemical tests. The biological species concept discriminates species based on their ability to mate. The application of molecular tools, such as DNA sequencing and phylogenetic analysis, to study diversity has greatly enhanced the resolution and added robustness to estimates of genetic diversity within various taxonomic groups.

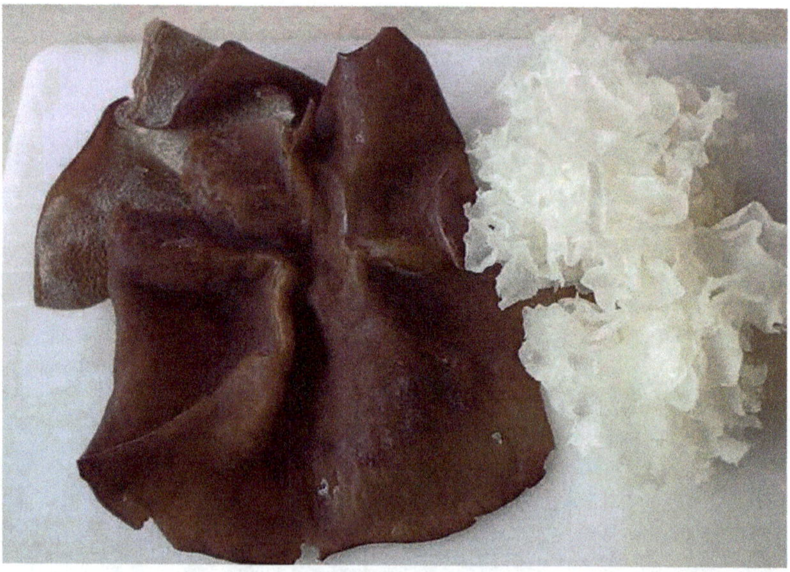

Two types of edible fungi

MYCOLOGY

Mycology is the branch of biology concerned with the systematic study of fungi, including their genetic and biochemical properties, their taxonomy, and their use to humans as a source of medicine, food, and psychotropic substances consumed for religious purposes, as well as their dangers, such as poisoning or infection. The field of phytopathology, the study of plant diseases, is closely related because many plant pathogens are fungi.

In 1729, Pier A. Micheli first published descriptions of fungi.

The use of fungi by humans dates back to prehistory; Ötzi the Iceman, a well-preserved mummy of a 5,300-year-old Neolithic man found frozen in the Austrian Alps, carried two species of polypore mushrooms that may have been used as tinder (*Fomes fomentarius*), or for medicinal purposes (*Piptoporus betulinus*). Ancient peoples have used fungi as food sources–often unknowingly–for millennia, in the preparation of leavened bread and fermented juices. Some of the oldest written records contain references to the destruction of crops that were probably caused by pathogenic fungi.

Mycology is a relatively new science that became systematic after the development of the microscope in the 17th century. Although fungal spores were first observed by Giambattista della Porta in 1588, the seminal work in the development of mycology is considered to be the publication of Pier Antonio Micheli's 1729 work *Nova plantarum genera*. Micheli not only observed spores but also showed that, under the proper conditions, they could be induced into growing into the same species of fungi from which they originated. Extending the use of the binomial system of nomenclature introduced by Carl Linnaeus in his *Species plantarum* (1753), the Dutch Christian Hendrik Persoon

(1761–1836) established the first classification of mushrooms with such skill so as to be considered a founder of modern mycology. Later, Elias Magnus Fries (1794–1878) further elaborated the classification of fungi, using spore color and various microscopic characteristics, methods still used by taxonomists today. Other notable early contributors to mycology in the 17th–19th and early 20th centuries include Miles Joseph Berkeley, August Carl Joseph Corda, Anton de Bary, the brothers Louis René and Charles Tulasne, Arthur H. R. Buller, Curtis G. Lloyd, and Pier Andrea Saccardo. The 20th century has seen a modernization of mycology that has come from advances in biochemistry, genetics, molecular biology, and biotechnology. The use of DNA sequencing technologies and phylogenetic analysis has provided new insights into fungal relationships and biodiversity, and has challenged traditional morphology-based groupings in fungal taxonomy.

MORPHOLOGY

Microscopic Structures

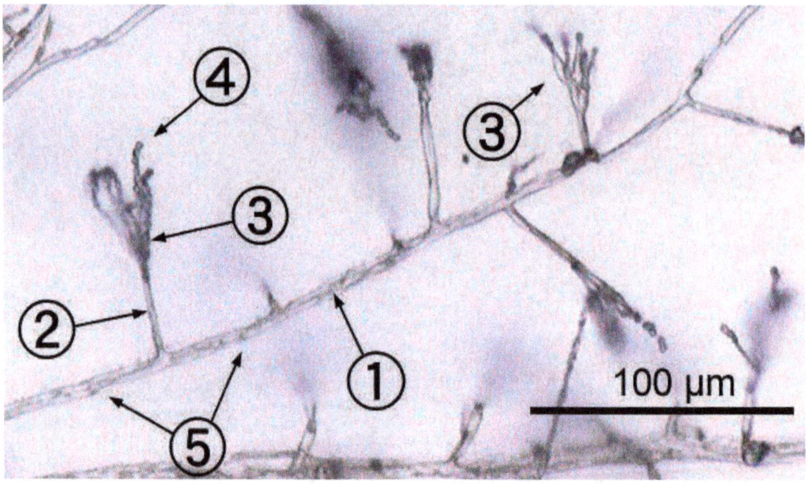

An environmental isolate of *Penicillium*

1. hypha
2. conidiophore
3. phialide
4. conidia
5. septa

Most fungi grow as hyphae, which are cylindrical, thread-like structures 2–10 µm in diameter and up to several centimeters in length. Hyphae grow at their tips (apices); new hyphae are typically formed by emergence of new tips along existing hyphae by a process called *branching*, or occasionally growing hyphal tips fork, giving rise to two parallel-growing hyphae. The combination of apical growth and branching/forking leads to the development of a mycelium, an interconnected network of hyphae. Hyphae can be either septate or coenocytic. Septate hyphae are divided into compartments separated by cross walls (internal cell walls, called septa, that are formed at right angles to the cell wall giving the hypha its shape), with each compartment containing one or more nuclei; coenocytic hyphae are not compartmentalized. Septa have pores that allow cytoplasm, organelles, and sometimes nuclei to pass through; an example is the dolipore septum in fungi of the phylum Basidiomycota. Coenocytic hyphae are in essence multinucleate supercells.

Many species have developed specialized hyphal structures for nutrient uptake from living hosts; examples include haustoria in plant-parasitic species of most fungal phyla, and arbuscules of several mycorrhizal fungi, which penetrate into the host cells to consume nutrients.

Although fungi are opisthokonts—a grouping of evolutionarily related organisms broadly characterized by a single posterior flagellum—all phyla except for the chytrids have lost their posterior flagella. Fungi are unusual among the eukaryotes in having a cell wall that, in addition to glucans (e.g., β-1,3-glucan) and other typical components, also contains the biopolymer chitin.

Macroscopic Structures

Armillaria solidipes

Fungal mycelia can become visible to the naked eye, for example, on various surfaces and substrates, such as damp walls and spoiled food, where they are commonly called molds. Mycelia grown on solid agar media in laboratory petri dishes are usually referred to as colonies. These colonies can exhibit growth shapes and colors (due to spores or pigmentation) that can be used as diagnostic features in the identification of species or groups. Some individual fungal colonies can reach extraordinary dimensions and ages as in the case of a clonal colony of *Armillaria solidipes*, which extends over an area of more than 900 ha (3.5 square miles), with an estimated age of nearly 9,000 years.

The apothecium—a specialized structure important in sexual reproduction in the ascomycetes—is a cup-shaped fruit body that holds the hymenium, a layer of tissue containing the spore-bearing cells. The fruit bodies of the basidiomycetes (basidiocarps) and some ascomycetes can sometimes grow very large, and many are well known as mushrooms.

GROWTH AND PHYSIOLOGY

Mold growth covering a decaying peach. The frames were taken approximately 12 hours apart over a period of six days.

The growth of fungi as hyphae on or in solid substrates or as single cells in aquatic environments is adapted for the efficient extraction of nutrients, because these growth forms have high surface area to volume ratios. Hyphae are specifically adapted for growth on solid surfaces, and to invade substrates and tissues. They can exert large penetrative mechanical forces; for example, the plant pathogen *Magnaporthe grisea* forms a structure called an appressorium that evolved to puncture plant tissues. The pressure generated by the appressorium, directed against the plant epidermis, can exceed 8 megapascals (1,200 psi). The filamentous fungus *Paecilomyces lilacinus* uses a similar structure to penetrate the eggs of nematodes.

The mechanical pressure exerted by the appressorium is generated from physiological processes that increase intracellular turgor by producing osmolytes such as glycerol. Adaptations such as these are complemented by hydrolytic enzymes secreted into the environment to digest large organic molecules—such as polysaccharides, proteins, and lipids—into smaller molecules that may then be absorbed as nutrients. The vast majority of filamentous fungi grow in a polar fashion—i.e., by extension into one direction—by elongation at the tip (apex) of the hypha. Other forms of fungal growth include intercalary extension (longitudinal expansion of hyphal compartments that are below the apex) as in the case of some endophytic fungi, or growth by volume expansion during the development of mushroom stipes and other large organs. Growth of fungi as multicellular structures consisting of somatic and reproductive cells—a feature independently evolved in animals and plants—has several functions, including the development of fruit bodies for dissemination of sexual spores (see above) and biofilms for substrate colonization and intercellular communication.

The fungi are traditionally considered heterotrophs, organisms that rely solely on carbon fixed by other organisms for metabolism. Fungi have evolved a high degree of metabolic versatility that allows them to use a diverse range of organic substrates for growth, including simple compounds such as nitrate, ammonia, acetate, or ethanol. In some species the pigment melanin may play a role in extracting energy from ionizing radiation, such as gamma radiation. This form of radiotrophic growth has been described for only a few species, the effects on growth rates are small, and the underlying biophysical and biochemical processes are not well known. This process might bear similarity to CO_2 fixation via visible light, but instead uses ionizing radiation as a source of energy.

REPRODUCTION

Polyporus squamosus

Fungal reproduction is complex, reflecting the differences in lifestyles and genetic makeup within this diverse kingdom of organisms. It is estimated that a third of all fungi reproduce using more than one method of propagation; for example, reproduction may occur in two well-differentiated stages within the life cycle of a species, the teleomorph and the anamorph. Environmental conditions trigger genetically determined developmental states that lead to the creation of specialized structures for sexual or asexual reproduction. These structures aid reproduction by efficiently dispersing spores or spore-containing propagules.

Asexual Reproduction

Asexual reproduction occurs via vegetative spores (conidia) or through mycelial fragmentation. Mycelial fragmentation occurs when a fungal mycelium separates into pieces, and each component grows into a separate mycelium. Mycelial fragmentation and vegetative spores maintain clonal populations adapted to a specific niche, and allow more rapid dispersal than sexual reproduction. The Fungi imperfecti (fungi lacking the perfect or sexual stage) or Deuteromycota comprise all the species that lack an observable sexual cycle.

Sexual Reproduction

Sexual reproduction with meiosis exists in all fungal phyla except Glomeromycota. It differs in many aspects from sexual reproduction in animals or plants. Differences also exist between fungal groups and can be used to discriminate species by morphological differences in sexual structures and reproductive strategies. Mating experiments between fungal isolates may identify species on the basis of biological species concepts. The major fungal groupings have initially been delineated based on the morphology of their sexual structures and spores; for example, the spore-containing structures, asci and basidia, can be used in the identification of ascomycetes and basidiomycetes, respectively. Some species may allow mating only between individuals of opposite mating type, whereas others can mate and sexually reproduce with any other individual or itself. Species of the former mating system are called heterothallic, and of the latter homothallic.

Most fungi have both a haploid and a diploid stage in their life cycles. In sexually reproducing fungi, compatible individuals may combine by fusing their hyphae together into an interconnected network; this process, anastomosis, is required for the initiation of the sexual cycle. Ascomycetes and basidiomycetes go through a dikaryotic stage, in which the nuclei inherited from the two parents do not combine immediately after cell fusion, but remain separate in the hyphal cells (see heterokaryosis).

In ascomycetes, dikaryotic hyphae of the hymenium (the spore-bearing tissue layer) form a characteristic *hook* at the hyphal septum. During cell division, formation of the hook ensures proper distribution of the newly divided nuclei into the apical and basal hyphal compartments. An ascus (plural *asci*) is then formed, in which karyogamy (nuclear fusion) occurs. Asci are embedded in an ascocarp, or fruiting body. Karyogamy in the asci is followed immediately by meiosis and the production of ascospores. After dispersal, the ascospores may germinate and form a new haploid mycelium.

Sexual reproduction in basidiomycetes is similar to that of the ascomycetes. Compatible haploid hyphae fuse to produce a dikaryotic mycelium. However, the dikaryotic phase is more extensive in the basidiomycetes, often also present in the vegetatively growing mycelium. A specialized anatomical structure, called a clamp connection, is formed at each hyphal septum. As with the structurally similar hook in the ascomycetes, the clamp connection in the basidiomycetes is required for controlled transfer of nuclei during cell division, to maintain the dikaryotic stage with two genetically different nuclei in each hyphal compartment. A basidiocarp is formed in which club-like structures known as basidia generate haploid basidiospores after karyogamy and meiosis. The most commonly known basidiocarps are mushrooms, but they may also take other forms (see Morphology section).

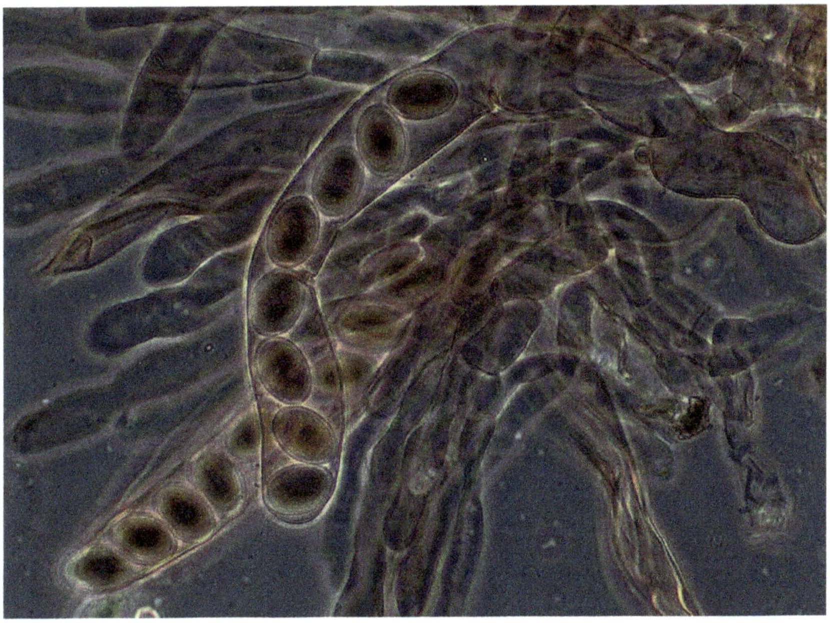

The 8-spore asci of *Morchella elata*, viewed with phase contrast microscopy

In glomeromycetes (formerly zygomycetes), haploid hyphae of two individuals fuse, forming a gametangium, a specialized cell structure that becomes a fertile gamete-producing cell. The gametangium develops into a zygospore, a thick-walled spore formed by the union of gametes. When the zygospore germinates, it undergoes meiosis, generating new haploid hyphae, which may then form asexual sporangiospores. These sporangiospores allow the fungus to rapidly disperse and germinate into new genetically identical haploid fungal mycelia.

Spore Dispersal

Both asexual and sexual spores or sporangiospores are often actively dispersed by forcible ejection from their reproductive structures. This ejection ensures exit of the spores from the reproductive structures as well as traveling through the air over long distances.

Specialized mechanical and physiological mechanisms, as well as spore surface structures (such as hydrophobins), enable efficient spore ejection. For example, the structure of the spore-bearing cells in some ascomycete species is such that the buildup of substances affecting cell volume and fluid balance enables the explosive discharge of spores into the air. The forcible discharge of single spores termed *ballistospores* involves formation of a small drop of water (Buller's drop), which upon contact with the spore leads to its projectile release with an initial acceleration of more than 10,000 g; the net result is that the spore is ejected 0.01–0.02 cm, sufficient distance for it to fall through the gills or pores into the air below. Other fungi, like the puffballs, rely on alternative mechanisms for spore release, such as external mechanical forces. The bird's nest fungi use the force of falling water drops to liberate the spores from cup-shaped fruiting bodies. Another strategy is seen in the stinkhorns, a group of fungi with lively colors and putrid odor that attract insects to disperse their spores.

The bird's nest fungus *Cyathus stercoreus*

Other Sexual Processes

Besides regular sexual reproduction with meiosis, certain fungi, such as those in the genera *Penicillium* and *Aspergillus*, may exchange genetic material via parasexual processes, initiated by anastomosis between hyphae and plasmogamy of fungal cells. The frequency and relative importance of parasexual events is unclear and may be lower than other sexual processes. It is known to play a role in intraspecific hybridization and is likely required for hybridization between species, which has been associated with major events in fungal evolution.

Evolution

In contrast to plants and animals, the early fossil record of the fungi is meager. Factors that likely contribute to the under-representation of fungal species among fossils include the nature of fungal fruiting bodies, which are soft, fleshy, and easily degradable tissues and the microscopic dimensions of most fungal structures, which therefore are not readily evident. Fungal fossils are difficult to distinguish from those of other microbes, and are most easily identified when they resemble extant fungi. Often recovered from a permineralized plant or animal host, these samples are typically studied by making thin-section preparations that can be examined with light microscopy or transmission electron microscopy. Researchers study compression fossils by dissolving the surrounding matrix with acid and then using light or scanning electron microscopy to examine surface details.

The earliest fossils possessing features typical of fungi date to the Proterozoic eon, some 1,430 million years ago (Ma); these multicellular benthic organisms had filamentous structures with septa, and were capable of anastomosis. More recent studies (2009) estimate the arrival of fungal organisms at about 760–1060 Ma on the basis of comparisons of the rate of evolution in closely related groups. For much of the Paleozoic Era (542–251 Ma), the fungi appear to have been aquatic and consisted of organisms similar to the extant chytrids in having flagellum-bearing spores. The evolutionary adaptation from an aquatic to a terrestrial lifestyle necessitated a diversification of ecological strategies for obtaining nutrients, including parasitism, saprobism, and the development of mutualistic relationships such as mycorrhiza and lichenization. Recent (2009) studies suggest that the ancestral ecological state of the Ascomycota was saprobism, and that independent lichenization events have occurred multiple times.

It is presumed that the fungi colonized the land during the Cambrian (542–488.3 Ma), long before land plants. Fossilized hyphae and spores recovered from the Ordovician of Wisconsin (460 Ma) resemble modern-day Glomerales, and existed at a time when the land flora likely consisted of only non-vascular bryophyte-like plants. Prototaxites, which was probably a fungus or lichen, would have been the tallest organism of the late Silurian. Fungal fossils do not become common and uncontroversial until the early Devonian (416–359.2 Ma), when they occur abundantly in the Rhynie chert, mostly as Zygomycota and Chytridiomycota. At about this same time, approximately 400 Ma, the Ascomycota and Basidiomycota diverged, and all modern classes of fungi were present by the Late Carboniferous (Pennsylvanian, 318.1–299 Ma).

Lichen-like fossils have been found in the Doushantuo Formation in southern China dating back to 635–551 Ma. Lichens formed a component of the early terrestrial ecosystems, and the estimated age of the oldest terrestrial lichen fossil is 400 Ma; this date corresponds to the age of the oldest known sporocarp fossil, a *Paleopyrenomycites* species found in the Rhynie Chert. The oldest fossil with microscopic features resembling modern-day basidiomycetes is *Palaeoancistrus*, found permineralized with a fern from the Pennsylvanian. Rare in the fossil record are the Homobasidiomycetes (a taxon roughly equivalent to the mushroom-producing species of the Agaricomycetes). Two amber-preserved specimens provide evidence that the earliest known mushroom-forming fungi (the extinct species *Archaeomarasmius leggetti*) appeared during the late Cretaceous, 90 Ma.

Some time after the Permian–Triassic extinction event (251.4 Ma), a fungal spike (originally thought to be an extraordinary abundance of fungal spores in sediments) formed, suggesting that fungi were the dominant life form at this time, representing nearly 100% of the available fossil record for this period. However, the relative proportion of fungal spores relative to spores formed by algal species is difficult to assess, the spike did not appear worldwide, and in many places it did not fall on the Permian–Triassic boundary.

TAXONOMY

Although commonly included in botany curricula and textbooks, fungi are more closely related to animals than to plants and are placed with the animals in the monophyletic group of opisthokonts. Analyses using molecular phylogenetics support a monophyletic origin of the Fungi. The taxonomy of the Fungi is in a state of constant flux, especially due to recent research based on DNA comparisons. These current phylogenetic analyses often overturn classifications based on older and sometimes less discriminative methods based on morphological features and biological species concepts obtained from experimental matings.

There is no unique generally accepted system at the higher taxonomic levels and there are frequent name changes at every level, from species upwards. Efforts among researchers are now underway to establish and encourage usage of a unified and more consistent nomenclature. Fungal species can also have multiple scientific names depending on their life cycle and mode (sexual or asexual) of reproduction. Web sites such as Index Fungorum and ITIS list current names of fungal species (with cross-references to older synonyms).

The 2007 classification of Kingdom Fungi is the result of a large-scale collaborative research effort involving dozens of mycologists and other scientists working on fungal taxonomy. It recognizes seven phyla, two of which—the Ascomycota and the Basidiomycota—are contained within a branch representing subkingdom Dikarya, the most species rich and familiar group, including all the mushrooms, most food-spoilage molds, most plant pathogenic fungi, and the beer, wine, and bread yeasts. The accompanying cladogram depicts the major fungal taxa and their relationship to opisthokont and unikont organisms, based on the work of Philippe Silar and "The Mycota: A Comprehensive Treatise on Fungi as Experimental Systems for Basic and Applied Research". The lengths of the branches are not proportional to evolutionary distances.

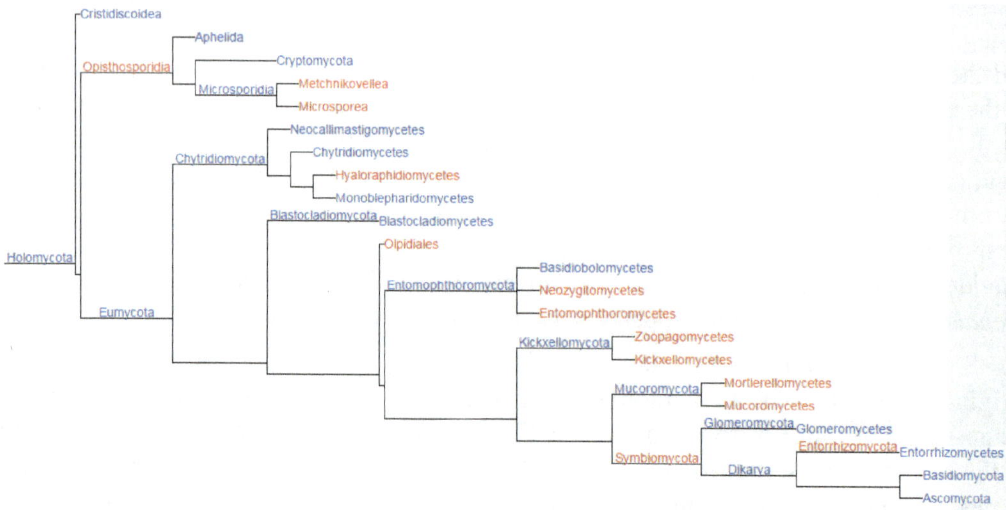

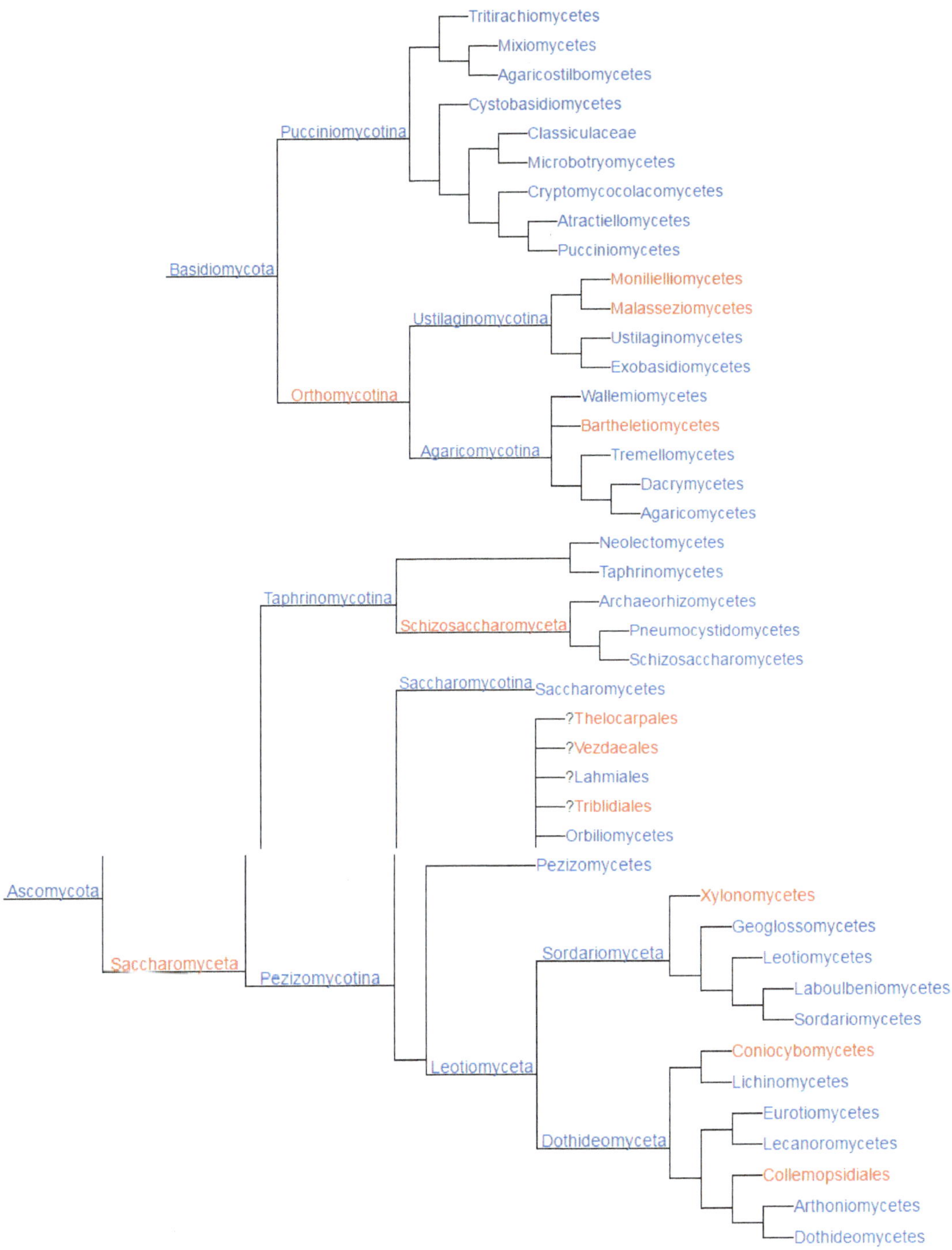

Taxonomic Groups

The major phyla (sometimes called divisions) of fungi have been classified mainly on the basis of characteristics of their sexual reproductive structures. Currently, seven phyla are proposed: Microsporidia, Chytridiomycota, Blastocladiomycota, Neocallimastigomycota, Glomeromycota, Ascomycota, and Basidiomycota.

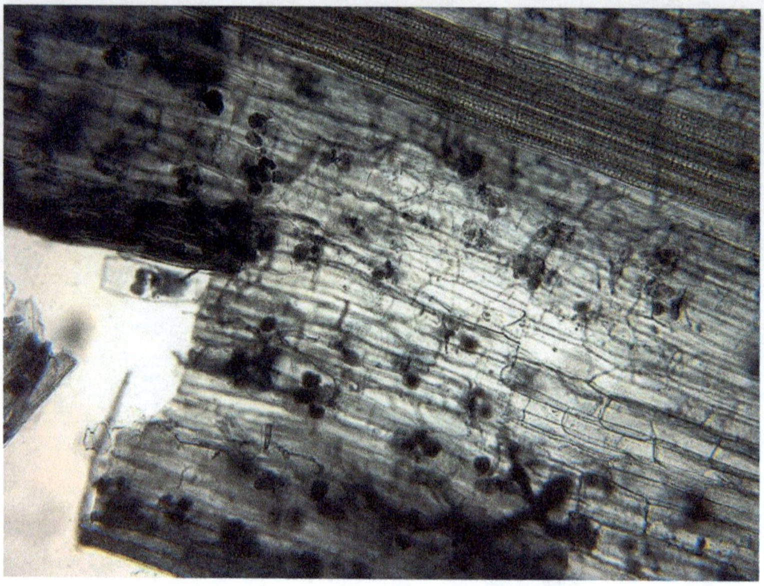

Arbuscular mycorrhiza seen under microscope. Flax root cortical cells containing paired arbuscules.

Phylogenetic analysis has demonstrated that the Microsporidia, unicellular parasites of animals and protists, are fairly recent and highly derived endobiotic fungi (living within the tissue of another species). One 2006 study concludes that the Microsporidia are a sister group to the true fungi; that is, they are each other's closest evolutionary relative. Hibbett and colleagues suggest that this analysis does not clash with their classification of the Fungi, and although the Microsporidia are elevated to phylum status, it is acknowledged that further analysis is required to clarify evolutionary relationships within this group.

The Chytridiomycota are commonly known as chytrids. These fungi are distributed worldwide. Chytrids and their close relatives Neocallimastigomycota and Blastocladiomycota (below) are the only fungi with active motility, producing zoospores that are capable of active movement through aqueous phases with a single flagellum, leading early taxonomists to classify them as protists. Molecular phylogenies, inferred from rRNA sequences in ribosomes, suggest that the Chytrids are a basal group divergent from the other fungal phyla, consisting of four major clades with suggestive evidence for paraphyly or possibly polyphyly.

The Blastocladiomycota were previously considered a taxonomic clade within the Chytridiomycota. Recent molecular data and ultrastructural characteristics, however, place the Blastocladiomycota as a sister clade to the Zygomycota, Glomeromycota, and Dikarya (Ascomycota and Basidiomycota). The blastocladiomycetes are saprotrophs, feeding on decomposing organic matter, and they are parasites of all eukaryotic groups. Unlike their close relatives, the chytrids, most of which exhibit zygotic meiosis, the blastocladiomycetes undergo sporic meiosis. The Neocallimastigomycota were earlier placed in the phylum Chytridomycota. Members of this small phylum are anaerobic organisms, living in the digestive system of larger herbivorous mammals and in other terrestrial and aquatic environments enriched in cellulose (e.g., domestic waste landfill sites). They lack mitochondria but contain hydrogenosomes of mitochondrial origin. As in the related chrytrids, neocallimastigomycetes form zoospores that are posteriorly uniflagellate or polyflagellate.

Members of the Glomeromycota form arbuscular mycorrhizae, a form of mutualist symbiosis wherein fungal hyphae invade plant root cells and both species benefit from the resulting increased supply of nutrients. All known Glomeromycota species reproduce asexually. The symbiotic association between the Glomeromycota and plants is ancient, with evidence dating to 400 million years ago. Formerly part of the Zygomycota (commonly known as 'sugar' and 'pin' molds), the Glomeromycota were elevated to phylum status in 2001 and now replace the older phylum Zygomycota. Fungi that were placed in the Zygomycota are now being reassigned to the Glomeromycota, or the subphyla incertae sedis Mucoromycotina, Kickxellomycotina, the Zoopagomycotina and the Entomophthoromycotina. Some well-known examples of fungi formerly in the Zygomycota include black bread mold (*Rhizopus stolonifer*), and *Pilobolus* species, capable of ejecting spores several meters through the air. Medically relevant genera include *Mucor*, *Rhizomucor*, and *Rhizopus*.

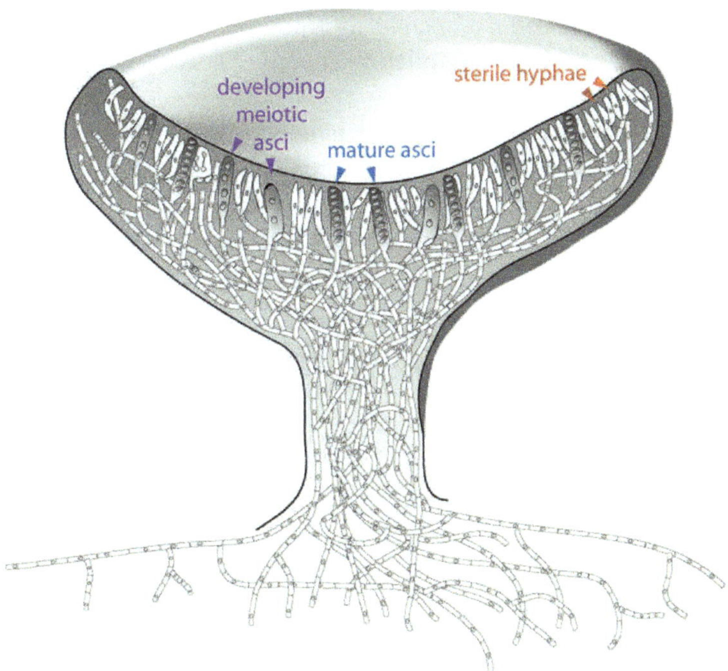

Diagram of an apothecium (the typical cup-like reproductive structure of Ascomycetes) showing sterile tissues as well as developing and mature asci.

The Ascomycota, commonly known as sac fungi or ascomycetes, constitute the largest taxonomic group within the Eumycota. These fungi form meiotic spores called ascospores, which are enclosed in a special sac-like structure called an ascus. This phylum includes morels, a few mushrooms and truffles, unicellular yeasts (e.g., of the genera *Saccharomyces*, *Kluyveromyces*, *Pichia*, and *Candida*), and many filamentous fungi living as saprotrophs, parasites, and mutualistic symbionts. Prominent and important genera of filamentous ascomycetes include *Aspergillus*, *Penicillium*, *Fusarium*, and *Claviceps*. Many ascomycete species have only been observed undergoing asexual reproduction (called anamorphic species), but analysis of molecular data has often been able to identify their closest teleomorphs in the Ascomycota. Because the products of meiosis are retained within the sac-like ascus, ascomycetes have been used for elucidating principles of genetics and heredity (e.g., *Neurospora crassa*).

Members of the Basidiomycota, commonly known as the club fungi or basidiomycetes, produce meiospores called basidiospores on club-like stalks called basidia. Most common mushrooms belong to this group, as well as rust and smut fungi, which are major pathogens of grains. Other important basidiomycetes include the maize pathogen *Ustilago maydis*, human commensal species of the genus *Malassezia*, and the opportunistic human pathogen, *Cryptococcus neoformans*.

Fungus-Like Organisms

Because of similarities in morphology and lifestyle, the slime molds (mycetozoans, plasmodiophorids, acrasids, *Fonticula* and labyrinthulids, now in Amoebozoa, Rhizaria, Excavata, Opisthokonta and Stramenopiles, respectively), water molds (oomycetes) and hyphochytrids (both Stramenopiles) were formerly classified in the kingdom Fungi, in groups like Mastigomycotina, Gymnomycota and Phycomycetes. The slime molds were studied also as protozoans, leading to a ambiregnal, duplicated taxonomy.

Unlike true fungi, the cell walls of oomycetes contain cellulose and lack chitin. Hyphochytrids have both chitin and cellulose. Slime molds lack a cell wall during the assimilative phase (except labyrinthulids, which have a wall of scales), and ingest nutrients by ingestion (phagocytosis, except labyrinthulids) rather than absorption (osmotrophy, as fungi, labyrinthulids, oomycetes and hyphochytrids). Neither water molds nor slime molds are closely related to the true fungi, and, therefore, taxonomists no longer group them in the kingdom Fungi. Nonetheless, studies of the oomycetes and myxomycetes are still often included in mycology textbooks and primary research literature.

The Eccrinales and Amoebidiales are opisthokont protists, previously thought to be zygomycete fungi. Other groups now in Opisthokonta (e.g., *Corallochytrium*, Ichthyosporea) were also at given time classified as fungi. The genus *Blastocystis*, now in Stramenopiles, was originally classified as a yeast. *Ellobiopsis*, now in Alveolata, was considered a chytrid. The bacteria were also included in fungi in some classifications, as the group Schizomycetes.

The Rozellida clade, including the "ex-chytrid" *Rozella*, is a genetically disparate group known mostly from environmental DNA sequences that is a sister group to fungi. Members of the group that have been isolated lack the chitinous cell wall that is characteristic of fungi.

The nucleariids, protists currently grouped in the Choanozoa (Opisthokonta), may be the next sister group to the eumycete clade, and as such could be included in an expanded fungal kingdom.

Many Actinomycetales (Actinobacteria), a group with many filamentous bacteria, were also long believed to be fungi.

Ecology

A pin mold decomposing a peach

Although often inconspicuous, fungi occur in every environment on Earth and play very important roles in most ecosystems. Along with bacteria, fungi are the major decomposers in most terrestrial (and some aquatic) ecosystems, and therefore play a critical role in biogeochemical cycles and in many food webs. As decomposers, they play an essential role in nutrient cycling, especially as saprotrophs and symbionts, degrading organic matter to inorganic molecules, which can then re-enter anabolic metabolic pathways in plants or other organisms.

Symbiosis

Many fungi have important symbiotic relationships with organisms from most if not all Kingdoms. These interactions can be mutualistic or antagonistic in nature, or in the case of commensal fungi are of no apparent benefit or detriment to the host.

With Plants

Mycorrhizal symbiosis between plants and fungi is one of the most well-known plant–fungus associations and is of significant importance for plant growth and persistence in many ecosystems; over 90% of all plant species engage in mycorrhizal relationships with fungi and are dependent upon this relationship for survival.

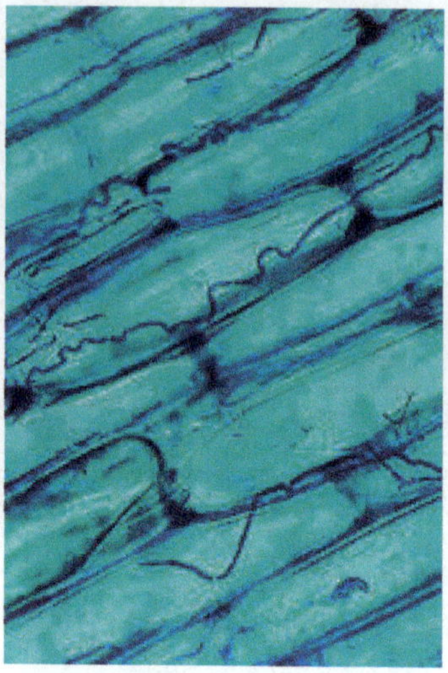

The dark filaments are hyphae of the endophytic fungus *Neotyphodium coenophialum* in the intercellular spaces of tall fescue leaf sheath tissue

The mycorrhizal symbiosis is ancient, dating to at least 400 million years ago. It often increases the plant's uptake of inorganic compounds, such as nitrate and phosphate from soils having low concentrations of these key plant nutrients. The fungal partners may also mediate plant-to-plant transfer of carbohydrates and other nutrients. Such mycorrhizal communities are called "common mycorrhizal networks". A special case of mycorrhiza is myco-heterotrophy, whereby the plant parasitizes the fungus, obtaining all of its nutrients from its fungal symbiont. Some fungal species inhabit the tissues inside roots, stems, and leaves, in which case they are called endophytes. Similar to mycorrhiza, endophytic colonization by fungi may benefit both symbionts; for example, endophytes of grasses impart to their host increased resistance to herbivores and other environmental stresses and receive food and shelter from the plant in return.

With Algae and Cyanobacteria

Lichens are a symbiotic relationship between fungi and algae or cyanobacteria. The algae partner in the relationship is referred to in lichen terminology as a photobiont. The fungi part of the relationship are composed mostly of various species of ascomycetes and a few basidiomycetes. Lichens occur in every ecosystem on all continents, play a key role in soil formation and the initiation of biological succession, and are the dominating life forms in extreme environments, including polar, alpine, and semiarid desert regions. They are able to grow on inhospitable surfaces, including bare soil, rocks, tree bark, wood, shells, barnacles and leaves. As in mycorrhizas, the photobiont provides sugars and other carbohydrates via photosynthesis to the fungus, while the fungus provides minerals and water to the photobiont. The functions of both symbiotic organisms are so closely intertwined that they function almost as a single organism; in most cases the resulting organism differs greatly from the individual components. Lichenization is a common mode of nutrition for fungi; around 20% of fungi—between 17,500 and

20,000 described species—are lichenized. Characteristics common to most lichens include obtaining organic carbon by photosynthesis, slow growth, small size, long life, long-lasting (seasonal) vegetative reproductive structures, mineral nutrition obtained largely from airborne sources, and greater tolerance of desiccation than most other photosynthetic organisms in the same habitat.

The lichen *Lobaria pulmonaria*, a symbiosis of fungal, algal, and cyanobacterial species

With Insects

Many insects also engage in mutualistic relationships with fungi. Several groups of ants cultivate fungi in the order Agaricales as their primary food source, while ambrosia beetles cultivate various species of fungi in the bark of trees that they infest. Likewise, females of several wood wasp species (genus *Sirex*) inject their eggs together with spores of the wood-rotting fungus *Amylostereum areolatum* into the sapwood of pine trees; the growth of the fungus provides ideal nutritional conditions for the development of the wasp larvae. At least one species of stingless bee has a relationship with a fungus in the genus *Monascus*, where the larvae consume and depend on fungus transferred from old to new nests. Termites on the African savannah are also known to cultivate fungi, and yeasts of the genera *Candida* and *Lachancea* inhabit the gut of a wide range of insects, including neuropterans, beetles, and cockroaches; it is not known whether these fungi benefit their hosts. Fungi ingrowing dead wood are essential for xylophagous insects (e.g. woodboring beetles). They deliver nutrients needed by xylophages to nutritionally scarce dead wood. Thanks to this nutritional enrichment the larvae of woodboring insect is able to grow and develop to adulthood.The larvae of many families of fungicolous flies, particularly those within the superfamily Sciaroidea such as the Mycetophilidae and some Keroplatidae feed on fungal fruiting bodies and sterile mycorrhizae.

As Pathogens and Parasites

The plant pathogen *Aecidium magellanicum* causes calafate rust, seen here on a *Berberis* shrub in Chile.

Many fungi are parasites on plants, animals (including humans), and other fungi. Serious pathogens of many cultivated plants causing extensive damage and losses to agriculture and forestry include the rice blast fungus *Magnaporthe oryzae*, tree pathogens such as *Ophiostoma ulmi* and *Ophiostoma novo-ulmi* causing Dutch elm disease, and *Cryphonectria parasitica* responsible for chestnut blight, and plant pathogens in the genera *Fusarium*, *Ustilago*, *Alternaria*, and *Cochliobolus*. Some carnivorous fungi, like *Paecilomyces lilacinus*, are predators of nematodes, which they capture using an array of specialized structures such as constricting rings or adhesive nets.

Some fungi can cause serious diseases in humans, several of which may be fatal if untreated. These include aspergillosis, candidiasis, coccidioidomycosis, cryptococcosis, histoplasmosis, mycetomas, and paracoccidioidomycosis. Furthermore, persons with immuno-deficiencies are particularly susceptible to disease by genera such as *Aspergillus*, *Candida*, *Cryptoccocus*, *Histoplasma*, and *Pneumocystis*. Other fungi can attack eyes, nails, hair, and especially skin, the so-called dermatophytic and keratinophilic fungi, and cause local infections such as ringworm and athlete's foot.Fungal spores are also a cause of allergies, and fungi from different taxonomic groups can evoke allergic reactions.

Mycotoxins

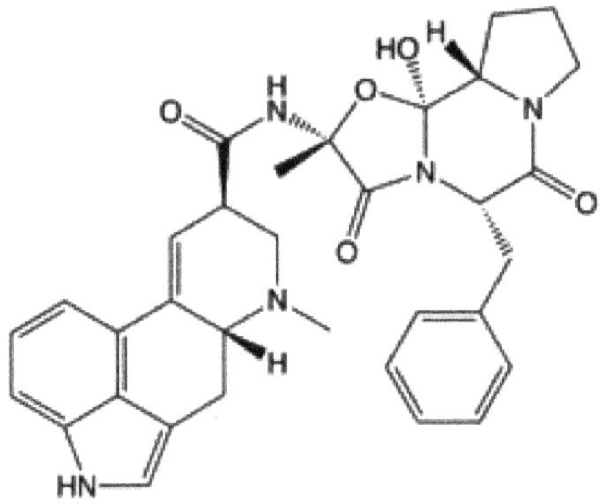

Ergotamine, a major mycotoxin produced by *Claviceps* species, which if ingested can cause gangrene, convulsions, and hallucinations

Many fungi produce biologically active compounds, several of which are toxic to animals or plants and are therefore called mycotoxins. Of particular relevance to humans are mycotoxins produced by molds causing food spoilage, and poisonous mushrooms (see above). Particularly infamous are the lethal amatoxins in some *Amanita* mushrooms, and ergot alkaloids, which have a long history of causing serious epidemics of ergotism (St Anthony's Fire) in people consuming rye or related cereals contaminated with sclerotia of the ergot fungus, *Claviceps purpurea*. Other notable mycotoxins include the aflatoxins, which are insidious liver toxins and highly carcinogenic metabolites produced by certain *Aspergillus* species often growing in or on grains and nuts consumed by humans, ochratoxins, patulin, and trichothecenes (e.g., T-2 mycotoxin) and fumonisins, which have significant impact on human food supplies or animal livestock.

Mycotoxins are secondary metabolites (or natural products), and research has established the existence of biochemical pathways solely for the purpose of producing mycotoxins and other natural products in fungi. Mycotoxins may provide fitness benefits in terms of physiological adaptation, competition with other microbes and fungi, and protection from consumption (fungivory).

Pathogenic Mechanisms

Ustilago maydis is a pathogenic plant fungus that causes smut disease in maize and teosinte. Plants have evolved efficient defense systems against pathogenic microbes such as *U. maydis*. A rapid defense reaction after pathogen attack is the oxidative burst where the plant produces reactive oxygen species at the site of the attempted invasion. *U. maydis* can respond to the oxidative burst with an oxidative stress response, regulated by the gene *YAP1*. The response protects *U. maydis* from the host defense, and is necessary for the pathogen's virulence. Furthermore, *U. maydis* has a well-established recombinational DNA repair system which acts during mitosis and meiosis. The system may assist the pathogen in surviving DNA damage arising from the host plant's oxidative defensive response to infection.

Cryptococcus neoformans is an encapsulated yeast that can live in both plants and animals. *C. neoformans* usually infects the lungs, where it is phagocytosed by alveolar macrophages. Some *C. neoformans* can survive inside macrophages, which appears to be the basis for latency, disseminated disease, and resistance to antifungal agents. One mechanism by which *C. neoformans* survives the hostile macrophage environment is by up-regulating the expression of genes involved in the oxidative stress response. Another mechanism involves meiosis. The majority of *C. neoformans* are mating "type a". Filaments of mating "type a ordinarily have haploid nuclei, but they can become diploid (perhaps by endoduplication or by stimulated nuclear fusion) to form blastospores. The diploid nuclei of blastospores can undergo meiosis, including recombination, to form haploid basidiospores that can be dispersed. This process is referred to as monokaryotic fruiting. this process requires a gene called *DMC1*, which is a conserved homologue of genes *recA* in bacteria and *RAD51* in eukaryotes, that mediates homologous chromosome pairing during meiosis and repair of DNA double-strand breaks. Thus, *C. neoformans* can undergo a meiosis, monokaryotic fruiting, that promotes recombinational repair in the oxidative, DNA damaging environment of the host macrophage, and the repair capability may contribute to its virulence.

Human Use

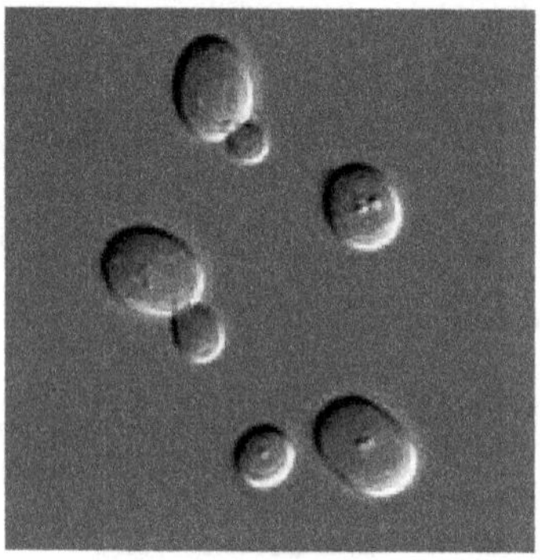

Saccharomyces cerevisiae cells shown with DIC microscopy

The human use of fungi for food preparation or preservation and other purposes is extensive and has a long history. Mushroom farming and mushroom gathering are large industries in many countries. The study of the historical uses and sociological impact of fungi is known as ethnomycology. Because of the capacity of this group to produce an enormous range of natural products with antimicrobial or other biological activities, many species have long been used or are being developed for industrial production of antibiotics, vitamins, and anti-cancer and cholesterol-lowering drugs. More recently, methods have been developed for genetic engineering of fungi, enabling metabolic engineering of fungal species. For example, genetic modification of yeast species—which are easy to grow at fast rates in large fermentation vessels—has opened up ways of pharmaceutical production that are potentially more efficient than production by the original source organisms.

Therapeutic Uses

Modern Chemotherapeutics

Many species produce metabolites that are major sources of pharmacologically active drugs. Particularly important are the antibiotics, including the penicillins, a structurally related group of β-lactam antibiotics that are synthesized from small peptides. Although naturally occurring penicillins such as penicillin G (produced by *Penicillium chrysogenum*) have a relatively narrow spectrum of biological activity, a wide range of other penicillins can be produced by chemical modification of the natural penicillins. Modern penicillins are semisynthetic compounds, obtained initially from fermentation cultures, but then structurally altered for specific desirable properties. Other antibiotics produced by fungi include: ciclosporin, commonly used as an immunosuppressant during transplant surgery; and fusidic acid, used to help control infection from methicillin-resistant *Staphylococcus aureus* bacteria. Widespread use of antibiotics for the treatment of bacterial diseases, such as tuberculosis, syphilis, leprosy, and others began in the early 20th century and continues to date. In nature, antibiotics of fungal or bacterial origin appear to play a dual role: at high concentrations they act as chemical defense against competition with other microorganisms in species-rich environments, such as the rhizosphere, and at low concentrations as quorum-sensing molecules for intra- or interspecies signaling. Other drugs produced by fungi include griseofulvin isolated from *Penicillium griseofulvum*, used to treat fungal infections, and statins (HMG-CoA reductase inhibitors), used to inhibit cholesterol synthesis. Examples of statins found in fungi include mevastatin from *Penicillium citrinum* and lovastatin from *Aspergillus terreus* and the oyster mushroom.

Traditional and Folk Medicine

The medicinal fungi *Ganoderma lucidum* (left) and *Ophiocordyceps sinensis* (right)

Certain mushrooms enjoy usage as therapeutics in folk medicines, such as Traditional Chinese medicine. Notable medicinal mushrooms with a well-documented history of use include *Agaricus subrufescens*, *Ganoderma lucidum*, and *Ophiocordyceps sinensis*. Research has identified compounds produced by these and other fungi that have inhibitory biological effects against viruses and cancer cells. Specific metabolites, such as polysaccharide-K, ergotamine, and β-lactam antibiotics, are routinely used in clinical medicine. The shiitake mushroom is a source of lentinan, a clinical drug approved for use in cancer treatments in several countries, including Japan. In Europe and Japan, polysaccharide-K (brand name Krestin), a chemical derived from *Trametes versicolor*, is an approved adjuvant for cancer therapy.

Cultured Foods

Baker's yeast or *Saccharomyces cerevisiae*, a unicellular fungus, is used to make bread and other wheat-based products, such as pizza dough and dumplings.Yeast species of the genus *Saccharomyces* are also used to produce alcoholic beverages through fermentation. Shoyu koji mold (*Aspergillus oryzae*) is an essential ingredient in brewing Shoyu (soy sauce) and sake, and the preparation of miso, while *Rhizopus* species are used for making tempeh. Several of these fungi are domesticated species that were bred or selected according to their capacity to ferment food without producing harmful mycotoxins (see below), which are produced by very closely related *Aspergilli*. Quorn, a meat substitute, is made from *Fusarium venenatum*.

Edible and Poisonous Species

Edible mushrooms are well-known examples of fungi. Many are commercially raised, but others must be harvested from the wild. *Agaricus bisporus*, sold as button mushrooms when small or Portobello mushrooms when larger, is a commonly eaten species, used in salads, soups, and many other dishes. Many Asian fungi are commercially grown and have increased in popularity in the West. They are often available fresh in grocery stores and markets, including straw mushrooms (*Volvariella volvacea*), oyster mushrooms (*Pleurotus ostreatus*), shiitakes (*Lentinula edodes*), and enokitake (*Flammulina* spp.).

Amanita phalloides accounts for the majority of fatal mushroom poisonings worldwide.

There are many more mushroom species that are harvested from the wild for personal consumption or commercial sale. Milk mushrooms, morels, chanterelles, truffles, black trumpets, and *porcini* mushrooms (*Boletus edulis*) (also known as king boletes) demand a high price on the market. They are often used in gourmet dishes.

Certain types of cheeses require inoculation of milk curds with fungal species that impart a unique flavor and texture to the cheese. Examples include the blue color in cheeses such as Stilton or Roquefort, which are made by inoculation with *Penicillium roqueforti*. Molds used in cheese production are non-toxic and are thus safe for human consumption; however, mycotoxins (e.g., aflatoxins, roquefortine C, patulin, or others) may accumulate because of growth of other fungi during cheese ripening or storage.

Many mushroom species are poisonous to humans, with toxicities ranging from slight digestive problems or allergic reactions as well as hallucinations to severe organ failures and death. Genera with mushrooms containing deadly toxins include *Conocybe*, *Galerina*, *Lepiota*, and, the most infamous, *Amanita*. The latter genus includes the destroying angel *(A. virosa)* and the death cap *(A. phalloides)*, the most common cause of deadly mushroom poisoning. The false morel (*Gyromitra esculenta*) is occasionally considered a delicacy when cooked, yet can be highly toxic when eaten raw. *Tricholoma equestre* was considered edible until it was implicated in serious poisonings causing rhabdomyolysis. Fly agaric mushrooms (*Amanita muscaria*) also cause occasional non-fatal poisonings, mostly as a result of ingestion for its hallucinogenic properties. Historically, fly agaric was used by different peoples in Europe and Asia and its present usage for religious or shamanic purposes is reported from some ethnic groups such as the Koryak people of north-eastern Siberia.

Stilton cheese veined with *Penicillium roqueforti*

As it is difficult to accurately identify a safe mushroom without proper training and knowledge, it is often advised to assume that a wild mushroom is poisonous and not to consume it.

Pest Control

In agriculture, fungi may be useful if they actively compete for nutrients and space with pathogenic microorganisms such as bacteria or other fungi via the competitive exclusion principle, or if they are parasites of these pathogens. For example, certain species may be used to eliminate or suppress the growth of harmful plant pathogens, such as insects, mites, weeds, nematodes, and other fungi that cause diseases of important crop plants. This has generated strong interest in practical applications that use these fungi in the biological control of these agricultural pests. Entomopathogenic fungi can be used as biopesticides, as they actively kill insects. Examples that have been used as biological insecticides are *Beauveria bassiana*, *Metarhizium* spp, *Hirsutella* spp, *Paecilomyces* (*Isaria*) spp, and *Lecanicillium lecanii*. Endophytic fungi of grasses of the genus *Neotyphodium*, such as *N. coenophialum*, produce alkaloids that are toxic to a range of invertebrate and vertebrate herbivores. These alkaloids protect grass plants from herbivory, but several endophyte alkaloids can poison grazing animals, such as cattle and sheep. Infecting cultivars of pasture or forage grasses with *Neotyphodium* endophytes is one approach being used in grass breeding programs; the fungal strains are selected for producing only alkaloids that increase resistance to herbivores such as insects, while being non-toxic to livestock.

Grasshoppers killed by *Beauveria bassiana*

Bioremediation

Certain fungi, in particular "white rot" fungi, can degrade insecticides, herbicides, pentachlorophenol, creosote, coal tars, and heavy fuels and turn them into carbon dioxide, water, and basic elements. Fungi have been shown to biomineralize uranium oxides, suggesting they may have application in the bioremediation of radioactively polluted sites.

Model Organisms

Several pivotal discoveries in biology were made by researchers using fungi as model organisms, that is, fungi that grow and sexually reproduce rapidly in the laboratory. For example, the one gene-one enzyme hypothesis was formulated by scientists using the bread mold *Neurospora crassa* to test their biochemical theories. Other important model fungi are *Aspergillus nidulans* and the yeasts *Saccaromyces cerevisiae* and *Schizosaccharomyces pombe*, each of which with a long history of use to investigate issues in eukaryotic cell biology and genetics, such as cell cycle regulation, chromatin structure, and gene regulation. Other fungal models have more recently emerged that address specific biological questions relevant to medicine, plant pathology, and industrial uses; examples include *Candida albicans*, a dimorphic, opportunistic human pathogen, *Magnaporthe grisea*, a plant pathogen, and *Pichia pastoris*, a yeast widely used for eukaryotic protein production.

Others

Fungi are used extensively to produce industrial chemicals like citric, gluconic, lactic, and malic acids, and industrial enzymes, such as lipases used in biological detergents, cellulases used in making cellulosic ethanol and stonewashed jeans, and amylases, invertases, proteases and xylanases. Several species, most notably *Psilocybin mushrooms* (colloquially known as *magic mushrooms*), are ingested for their psychedelic properties, both recreationally and religiously.

REFERENCES

1. Moore RT (1980). "Taxonomic proposals for the classification of marine yeasts and other yeast-like fungi including the smuts". Botanica Marina. 23: 361–373.
2. The classification system presented here is based on the 2007 phylogenetic study by Hibbett et al.
3. "Fungus". Oxford Dictionaries. Retrieved 26 February 2011.
4. Simpson DP (1979). Cassell's Latin Dictionary (5 ed.). London, UK: Cassell Ltd. p. 883. ISBN 0-304-52257-0.
5. Ainsworth, p. 2.
6. Mitzka W, ed. (1960). Etymologisches Wörterbuch der deutschen Sprache. Berlin: Walter de Gruyter.
7. Alexopoulos et al., p. 1.
8. "LIAS Glossary". Retrieved 14 August 2013.
9. Bruns T (October 2006). "Evolutionary biology: a kingdom revised". Nature. 443 (7113): 758–61. Bibcode:2006Natur.443..758B. doi:10.1038/443758a. PMID 17051197.
10. Baldauf SL, Palmer JD (December 1993). "Animals and fungi are each other's closest relatives: congruent evidence from multiple proteins". Proceedings of the National Academy of Sciences of the United States of America. 90 (24): 11558–62. Bibcode:1993PNAS...9011558B. doi:10.1073/pnas.90.24.11558. PMC 48023 . PMID 8265589.
11. Deacon, p. 4.
12. Deacon, pp. 128–129.
13. Alexopoulos et al., pp. 28–33.
14. Alexopoulos et al., pp. 31–32.
15. Shoji JY, Arioka M, Kitamoto K (2006). "Possible involvement of pleiomorphic vacuolar networks in nutrient recycling in filamentous fungi". Autophagy. 2 (3): 226–7. doi:10.4161/auto.2695. PMID 16874107.
16. Deacon, p. 58.
17. Zabriskie TM, Jackson MD (February 2000). "Lysine biosynthesis and metabolism in fungi". Natural Product Reports. 17 (1): 85–97. doi:10.1039/a801345d. PMID 10714900.

18. Xu H, Andi B, Qian J, West AH, Cook PF (2006). "The alpha-aminoadipate pathway for lysine biosynthesis in fungi". Cell Biochemistry and Biophysics. 46 (1): 43–64. doi:10.1385/CBB:46:1:43. PMID 16943623.
19. Alexopoulos et al., pp. 27–28.
20. Alexopoulos et al., p. 685.
21. Alexopoulos et al., p. 30.
22. Desjardin DE, Perry BA, Lodge DJ, Stevani CV, Nagasawa E (2010). "Luminescent Mycena: new and noteworthy species". Mycologia. 102 (2): 459–77. doi:10.3852/09-197. PMID 20361513.
23. Alexopoulos et al., pp. 32–33.
24. Bowman SM, Free SJ (August 2006). "The structure and synthesis of the fungal cell wall". BioEssays. 28 (8): 799–808. doi:10.1002/bies.20441. PMID 16927300.
25. Alexopoulos et al., p. 33.
26. Mihail JD, Bruhn JN (November 2005). "Foraging behaviour of Armillaria rhizomorph systems". Mycological Research. 109 (Pt 11): 1195–207. doi:10.1017/S0953756205003606. PMID 16279413.
27. Keller NP, Turner G, Bennett JW (December 2005). "Fungal secondary metabolism - from biochemistry to genomics". Nature Reviews. Microbiology. 3 (12): 937–47. doi:10.1038/nrmicro1286. PMID 16322742.
28. Wu S, Schalk M, Clark A, Miles RB, Coates R, Chappell J (November 2006). "Redirection of cytosolic or plastidic isoprenoid precursors elevates terpene production in plants". Nature Biotechnology. 24 (11): 1441–7. doi:10.1038/nbt1251. PMID 17057703.
29. Tudzynski B (March 2005). "Gibberellin biosynthesis in fungi: genes, enzymes, evolution, and impact on biotechnology". Applied Microbiology and Biotechnology. 66 (6): 597–611. doi:10.1007/s00253-004-1805-1. PMID 15578178.
30. Vaupotic T, Veranic P, Jenoe P, Plemenitas A (June 2008). "Mitochondrial mediation of environmental osmolytes discrimination during osmoadaptation in the extremely halotolerant black yeast Hortaea werneckii". Fungal Genetics and Biology. 45 (6): 994–1007. doi:10.1016/j.fgb.2008.01.006. PMID 18343697.
31. Dadachova E, Bryan RA, Huang X, Moadel T, Schweitzer AD, Aisen P, Nosanchuk JD, Casadevall A (2007). "Ionizing radiation changes the electronic properties of melanin and enhances the growth of melanized fungi". PloS One. 2 (5): e457. Bibcode:2007PLoSO...2..457D. doi:10.1371/journal.pone.0000457. PMC 1866175 . PMID 17520016.
32. Raghukumar C, Raghukumar S (1998). "Barotolerance of fungi isolated from deep-sea sediments of the Indian Ocean". Aquatic Microbial Ecology. 15 (2): 153–163. doi:10.3354/ame015153.
33. Sancho LG, de la Torre R, Horneck G, Ascaso C, de Los Rios A, Pintado A, Wierzchos J, Schuster M (June 2007). "Lichens survive in space: results from the 2005 LICHENS experiment". Astrobiology. 7 (3): 443–54. Bibcode:2007AsBio...7..443S. doi:10.1089/ast.2006.0046. PMID 17630840.

34. Brem FM, Lips KR (September 2008). "Batrachochytrium dendrobatidis infection patterns among Panamanian amphibian species, habitats and elevations during epizootic and enzootic stages". Diseases of Aquatic Organisms. 81 (3): 189–202. doi:10.3354/dao01960. PMID 18998584.

35. Le Calvez T, Burgaud G, Mahé S, Barbier G, Vandenkoornhuyse P (October 2009). "Fungal diversity in deep-sea hydrothermal ecosystems". Applied and Environmental Microbiology. 75 (20): 6415–21. doi:10.1128/AEM.00653-09. PMC 2765129 . PMID 19633124.

36. This estimation is determined by combining the species count for each phyla, based on values obtained from the 10th edition of the Dictionary of the Fungi (Kirk et al., 2008): Ascomycota, 64163 species (p. 55); Basidiomycota, 31515 (p. 78); Blastocladiomycota, 179 (p. 94); Chytridiomycota, 706 (p. 142); Glomeromycota, 169 (p. 287); Microsporidia, >1300 (p. 427); Neocallimastigomycota, 20 (p. 463).

37. Mueller GM, Schmit JP (2006). "Fungal biodiversity: what do we know? What can we predict?". Biodiversity and Conservation. 16: 1–5. doi:10.1007/s10531-006-9117-7.

38. Hawksworth DL (2006). "The fungal dimension of biodiversity: magnitude, significance, and conservation". Mycological Research. 95 (6): 641–655. doi:10.1016/S0953-7562(09)80810-1.

39. Blackwell M (March 2011). "The fungi: 1, 2, 3 ... 5.1 million species?". American Journal of Botany. 98 (3): 426–38. doi:10.3732/ajb.1000298. PMID 21613136.

40. Kirk et al., p. 489.

41. Hibbett DS, Binder M, Bischoff JF, Blackwell M, Cannon PF, Eriksson OE, et al. (May 2007). "A higher-level phylogenetic classification of the Fungi" (PDF). Mycological Research. 111 (Pt 5): 509–47. doi:10.1016/j.mycres.2007.03.004. PMID 17572334.

42. According to one 2001 estimate, some 10,000 fungal diseases are known. Struck C (2006). "Infection strategies of plant parasitic fungi". In Cooke BM, Jones DG, Kaye B. The Epidemiology of Plant Diseases. Berlin, Germany: Springer. p. 117. ISBN 1-4020-4580-8.

43. Peintner U, Pöder R, Pümpel T (1998). "The Iceman's fungi". Mycological Research. 102 (10): 1153–1162. doi:10.1017/S0953756298006546.

44. Ainsworth, p. 1.

45. Alexopoulos et al., pp. 1–2.

46. Ainsworth, p. 18.

47. Hawksworth DL (September 2006). "Pandora's mycological box: molecular sequences vs. morphology in understanding fungal relationships and biodiversity". Revista Iberoamericana de MicologíA. 23 (3): 127–33. doi:10.1016/S1130-1406(06)70031-6. PMID 17196017.

48. Harris SD (2008). "Branching of fungal hyphae: regulation, mechanisms and comparison with other branching systems". Mycologia. 100 (6): 823–32. doi:10.3852/08-177. PMID 19202837.

49. Deacon, p. 51.
50. Deacon, p. 57.
51. Chang S-T, Miles PG (2004). Mushrooms: Cultivation, Nutritional Value, Medicinal Effect and Environmental Impact. Boca Raton, Florida: CRC Press. ISBN 0-8493-1043-1.
52. Parniske M (October 2008). "Arbuscular mycorrhiza: the mother of plant root endosymbioses". Nature Reviews. Microbiology. 6 (10): 763–75. doi:10.1038/nrmicro1987. PMID 18794914.
53. Steenkamp ET, Wright J, Baldauf SL (January 2006). "The protistan origins of animals and fungi". Molecular Biology and Evolution. 23 (1): 93–106. doi:10.1093/molbev/msj011. PMID 16151185.
54. Stevens DA, Ichinomiya M, Koshi Y, Horiuchi H (September 2006). "Escape of Candida from caspofungin inhibition at concentrations above the MIC (paradoxical effect) accomplished by increased cell wall chitin; evidence for beta-1,6-glucan synthesis inhibition by caspofungin". Antimicrobial Agents and Chemotherapy. 50 (9): 3160–1. doi:10.1128/AAC.00563-06. PMC 1563524 . PMID 16940118.
55. Hanson, pp. 127–141.
56. Ferguson BA, Dreisbach TA, Parks CG, Filip GM, Schmitt CL (2003). "Coarse-scale population structure of pathogenic Armillaria species in a mixed-conifer forest in the Blue Mountains of northeast Oregon". Canadian Journal of Forest Research. 33 (4): 612–623. doi:10.1139/x03-065.
57. Alexopoulos et al., pp. 204–205.
58. Moss ST (1986). The Biology of Marine Fungi. Cambridge, UK: Cambridge University Press. p. 76. ISBN 0-521-30899-2.
59. Peñalva MA, Arst HN (September 2002). "Regulation of gene expression by ambient pH in filamentous fungi and yeasts". Microbiology and Molecular Biology Reviews. 66 (3): 426–46, table of contents. doi:10.1128/MMBR.66.3.426-446.2002. PMC 120796 . PMID 12208998.
60. Howard RJ, Ferrari MA, Roach DH, Money NP (December 1991). "Penetration of hard substrates by a fungus employing enormous turgor pressures". Proceedings of the National Academy of Sciences of the United States of America. 88 (24): 11281–4. Bibcode:1991PNAS...8811281H. doi:10.1073/pnas.88.24.11281. PMC 53118 . PMID 1837147.
61. Money NP (1998). "Mechanics of invasive fungal growth and the significance of turgor in plant infection". Molecular Genetics of Host-Specific Toxins in Plant Disease: Proceedings of the 3rd Tottori International Symposium on Host-Specific Toxins, Daisen, Tottori, Japan, August 24–29, 1997. Netherlands: Kluwer Academic Publishers. pp. 261–271. ISBN 0-7923-4981-4.
62. Wang ZY, Jenkinson JM, Holcombe LJ, Soanes DM, Veneault-Fourrey C, Bhambra GK, Talbot NJ (April 2005). "The molecular biology of appressorium turgor generation by the rice blast fungus Magnaporthe grisea". Biochemical Society Transactions. 33 (Pt 2): 384–8. doi:10.1042/BST0330384. PMID 15787612.

63. Pereira JL, Noronha EF, Miller RN, Franco OL (June 2007). "Novel insights in the use of hydrolytic enzymes secreted by fungi with biotechnological potential". Letters in Applied Microbiology. 44 (6): 573–81. doi:10.1111/j.1472-765X.2007.02151.x. PMID 17576216.
64. Schaller M, Borelli C, Korting HC, Hube B (November 2005). "Hydrolytic enzymes as virulence factors of Candida albicans". Mycoses. 48 (6): 365–77. doi:10.1111/j.1439-0507.2005.01165.x. PMID 16262871.
65. Farrar JF (October 1985). "Carbohydrate metabolism in biotrophic plant pathogens". Microbiological Sciences. 2 (10): 314–7. PMID 3939987.
66. Fischer R, Zekert N, Takeshita N (May 2008). "Polarized growth in fungi--interplay between the cytoskeleton, positional markers and membrane domains". Molecular Microbiology. 68 (4): 813–26. doi:10.1111/j.1365-2958.2008.06193.x. PMID 18399939.
67. Christensen MJ, Bennett RJ, Ansari HA, Koga H, Johnson RD, Bryan GT, Simpson WR, Koolaard JP, Nickless EM, Voisey CR (February 2008). "Epichloë endophytes grow by intercalary hyphal extension in elongating grass leaves". Fungal Genetics and Biology. 45 (2): 84–93. doi:10.1016/j.fgb.2007.07.013. PMID 17919950.
68. Money NP (October 2002). "Mushroom stem cells". BioEssays. 24 (10): 949–52. doi:10.1002/bies.10160. PMID 12325127.
69. Willensdorfer M (February 2009). "On the evolution of differentiated multicellularity". Evolution; International Journal of Organic Evolution. 63 (2): 306–23. doi:10.1111/j.1558-5646.2008.00541.x. PMID 19154376.
70. Daniels KJ, Srikantha T, Lockhart SR, Pujol C, Soll DR (May 2006). "Opaque cells signal white cells to form biofilms in Candida albicans". The EMBO Journal. 25 (10): 2240–52. doi:10.1038/sj.emboj.7601099. PMC 1462973 . PMID 16628217.
71. Marzluf GA (September 1981). "Regulation of nitrogen metabolism and gene expression in fungi". Microbiological Reviews. 45 (3): 437–61. PMC 281519 . PMID 6117784.
72. Hynes MJ (1994). "Regulatory circuits of the amdS gene of Aspergillus nidulans". Antonie Van Leeuwenhoek. 65 (3): 179–82. doi:10.1007/BF00871944. PMID 7847883.
73. Dadachova E, Casadevall A (December 2008). "Ionizing radiation: how fungi cope, adapt, and exploit with the help of melanin". Current Opinion in Microbiology. 11 (6): 525–31. doi:10.1016/j.mib.2008.09.013. PMC 2677413 . PMID 18848901.
74. Alexopoulos et al., pp. 48–56.
75. Kirk et al., p. 633.
76. Heitman J (September 2006). "Sexual reproduction and the evolution of microbial pathogens". Current Biology. 16 (17): R711–25. doi:10.1016/j.cub.2006.07.064. PMID 16950098.
77. Alcamo IE, Pommerville J (2004). Alcamo's Fundamentals of Microbiology.

Boston, Massachusetts: Jones and Bartlett. p. 590. ISBN 0-7637-0067-3.

78. Redecker D, Raab P (2006). "Phylogeny of the glomeromycota (arbuscular mycorrhizal fungi): recent developments and new gene markers". Mycologia. 98 (6): 885–95. doi:10.3852/mycologia.98.6.885. PMID 17486965.
79. Guarro J, Stchigel AM (July 1999). "Developments in fungal taxonomy". Clinical Microbiology Reviews. 12 (3): 454–500. PMC 100249 . PMID 10398676.
80. Taylor JW, Jacobson DJ, Kroken S, Kasuga T, Geiser DM, Hibbett DS, Fisher MC (October 2000). "Phylogenetic species recognition and species concepts in fungi". Fungal Genetics and Biology. 31 (1): 21–32. doi:10.1006/fgbi.2000.1228. PMID 11118132.
81. Metzenberg RL, Glass NL (February 1990). "Mating type and mating strategies in Neurospora". BioEssays. 12 (2): 53–9. doi:10.1002/bies.950120202. PMID 2140508.
82. Jennings and Lysek, pp. 107–114.
83. Deacon, p. 31.
84. Alexopoulos et al., pp. 492–493.
85. Jennings and Lysek, p. 142.
86. Deacon, pp. 21–24.
87. Linder MB, Szilvay GR, Nakari-Setälä T, Penttilä ME (November 2005). "Hydrophobins: the protein-amphiphiles of filamentous fungi". FEMS Microbiology Reviews. 29 (5): 877–96. doi:10.1016/j.femsre.2005.01.004. PMID 16219510.
88. Trail F (November 2007). "Fungal cannons: explosive spore discharge in the Ascomycota". FEMS Microbiology Letters. 276 (1): 12–8. doi:10.1111/j.1574-6968.2007.00900.x. PMID 17784861.
89. Pringle A, Patek SN, Fischer M, Stolze J, Money NP (2005). "The captured launch of a ballistospore". Mycologia. 97 (4): 866–71. doi:10.3852/mycologia.97.4.866. PMID 16457355.
90. Kirk et al., p. 495.
91. Brodie HJ (1975). The Bird's Nest Fungi. Toronto, Ontario: University of Toronto Press. p. 80. ISBN 0-8020-5307-6.
92. Alexopoulos et al., p. 545.
93. Jennings and Lysek, pp. 114–115.
94. Furlaneto MC, Pizzirani-Kleiner AA (January 1992). "Intraspecific hybridisation of Trichoderma pseudokoningii by anastomosis and by protoplast fusion". FEMS Microbiology Letters. 69 (2): 191–5. doi:10.1111/j.1574-6968.1992.tb05150.x. PMID 1537549.
95. Schardl CL, Craven KD (November 2003). "Interspecific hybridization in plant-associated fungi and oomycetes: a review". Molecular Ecology. 12 (11): 2861–73. doi:10.1046/j.1365-294X.2003.01965.x. PMID 14629368.
96. Donoghue MJ, Cracraft J (2004). Assembling the Tree of Life. Oxford

(Oxfordshire), UK: Oxford University Press. p. 187. ISBN 0-19-517234-5.

97. Taylor and Taylor, p. 19.
98. Taylor and Taylor, pp. 7–12.
99. Butterfield NJ (2005). "Probable Proterozoic fungi". Paleobiology. 31 (1): 165–182. doi:10.1666/0094-8373(2005)031<0165:PPF>2.0.CO;2.
100. Lücking R, Huhndorf S, Pfister DH, Plata ER, Lumbsch HT (2009). "Fungi evolved right on track". Mycologia. 101 (6): 810–22. doi:10.3852/09-016. PMID 19927746.
101. James TY, Kauff F, Schoch CL, Matheny PB, Hofstetter V, Cox CJ, et al. (October 2006). "Reconstructing the early evolution of Fungi using a six-gene phylogeny". Nature. 443 (7113): 818–22. Bibcode:2006Natur.443..818J. doi:10.1038/nature05110. PMID 17051209.
102. Taylor and Taylor, pp. 84–94 and 106–107.
103. Schoch CL, Sung GH, López-Giráldez F, Townsend JP, Miadlikowska J, Hofstetter V, et al. (April 2009). "The Ascomycota tree of life: a phylum-wide phylogeny clarifies the origin and evolution of fundamental reproductive and ecological traits". Systematic Biology. 58 (2): 224–39. doi:10.1093/sysbio/syp020. PMID 20525580.
104. Brundrett MC (2002). "Coevolution of roots and mycorrhizas of land plants". New Phytologist. 154 (2): 275–304. doi:10.1046/j.1469-8137.2002.00397.x.
105. Redecker D, Kodner R, Graham LE (September 2000). "Glomalean fungi from the Ordovician". Science. 289 (5486): 1920–1. Bibcode:2000Sci...289.1920R. doi:10.1126/science.289.5486.1920. PMID 10988069.
106. Taylor TN, Taylor EL (1996). "The distribution and interactions of some Paleozoic fungi". Review of Palaeobotany and Palynology. 95 (1–4): 83–94. doi:10.1016/S0034-6667(96)00029-2.
107. Dotzler N, Walker C, Krings M, Hass H, Kerp H, Taylor TN, Agerer R (2009). "Acaulosporoid glomeromycotan spores with a germination shield from the 400-million-year-old Rhynie chert". Mycological Progress. 8 (1): 9–18. doi:10.1007/s11557-008-0573-1.
108. Taylor JW, Berbee ML (2006). "Dating divergences in the Fungal Tree of Life: review and new analyses". Mycologia. 98 (6): 838–49. doi:10.3852/mycologia.98.6.838. PMID 17486961.
109. Blackwell M, Vilgalys R, James TY, Taylor JW (2009). "Fungi. Eumycota: mushrooms, sac fungi, yeast, molds, rusts, smuts, etc.". Tree of Life Web Project. Retrieved 25 April 2009.
110. Yuan X, Xiao S, Taylor TN (May 2005). "Lichen-like symbiosis 600 million years ago". Science. 308 (5724): 1017–20. Bibcode:2005Sci...308.1017Y. doi:10.1126/science.1111347. PMID 15890881.
111. Karatygin IV, Snigirevskaya NS, Vikulin SV (2009). "The most ancient terrestrial lichen Winfrenatia reticulata: A new find and new interpretation". Paleontological Journal. 43 (1): 107–114. doi:10.1134/S0031030109010110.

112. Taylor TN, Hass H, Kerp H, Krings M, Hanlin RT (2005). "Perithecial ascomycetes from the 400 million year old Rhynie chert: an example of ancestral polymorphism". Mycologia. 97 (1): 269–85. doi:10.3852/mycologia.97.1.269. PMID 16389979.

113. Dennis RL (1970). "A Middle Pennsylvanian basidiomycete mycelium with clamp connections". Mycologia. 62 (3): 578–584. doi:10.2307/3757529. JSTOR 3757529.

114. Hibbett DS, Grimaldi D, Donoghue MJ (1995). "Cretaceous mushrooms in amber". Nature. 377 (6549): 487. Bibcode:1995Natur.377..487H. doi:10.1038/377487a0.

115. Hibbett DS, Grimaldi D, Donoghue MJ (1997). "Fossil mushrooms from Miocene and Cretaceous ambers and the evolution of homobasidiomycetes". American Journal of Botany. 84 (7): 981–991. doi:10.2307/2446289. JSTOR 2446289.

116. Eshet Y, Rampino MR, Visscher H (1995). "Fungal event and palynological record of ecological crisis and recovery across the Permian-Triassic boundary". Geology. 23 (1): 967–970. Bibcode:1995Geo....23..967E. doi:10.1130/0091-7613(1995)023<0967:FEAPRO>2.3.CO;2.

117. Foster CB, Stephenson MH, Marshall C, Logan GA, Greenwood PF (2002). "A revision of Reduviasporonites Wilson 1962: description, illustration, comparison and biological affinities". Palynology. 26 (1): 35–58. doi:10.2113/0260035.

118. López-Gómez J, Taylor EL (2005). "Permian-Triassic transition in Spain: a multidisciplinary approach". Palaeogeography, Palaeoclimatology, Palaeoecology. 229 (1–2): 1–2. doi:10.1016/j.palaeo.2005.06.028.

119. Looy CV, Twitchett RJ, Dilcher DL, Van Konijnenburg-Van Cittert JH, Visscher H (July 2001). "Life in the end-Permian dead zone". Proceedings of the National Academy of Sciences of the United States of America. 98 (14): 7879–83. Bibcode:2001PNAS...98.7879L. doi:10.1073/pnas.131218098. PMC 35436 . PMID 11427710. See image 2

120. Ward PD, Botha J, Buick R, De Kock MO, Erwin DH, Garrison GH, Kirschvink JL, Smith R (February 2005). "Abrupt and gradual extinction among Late Permian land vertebrates in the Karoo basin, South Africa". Science. 307 (5710): 709–14. Bibcode:2005Sci...307..709W. doi:10.1126/science.1107068. PMID 15661973.

121. Shalchian-Tabrizi K, Minge MA, Espelund M, Orr R, Ruden T, Jakobsen KS, Cavalier-Smith T (2008). "Multigene phylogeny of choanozoa and the origin of animals". PloS One. 3 (5): e2098. Bibcode:2008PLoSO...3.2098S. doi:10.1371/journal.pone.0002098. PMC 2346548 . PMID 18461162.

122. See Palaeos Fungi: Fungi for an introduction to fungal taxonomy, including recent controversies. archive

123. Celio GJ, Padamsee M, Dentinger BT, Bauer R, McLaughlin DJ (2006). "Assembling the Fungal Tree of Life: constructing the structural and biochemical database". Mycologia. 98 (6): 850–9. doi:10.3852/mycologia.98.6.850. PMID 17486962.

124. Silar P (2016). "Protistes Eucaryotes: Origine, Evolution et Biologie des Microbes Eucaryotes". HAL: 462. ISBN 978-2-9555841-0-1.

125. Esser K (2014). The Mycota VII A: Systematics and Evolution (2nd ed.). Springer. p. 461. ISBN 978-3-642-55317-2.

126. Gill EE, Fast NM (June 2006). "Assessing the microsporidia-fungi relationship: Combined phylogenetic analysis of eight genes". Gene. 375: 103–9. doi:10.1016/j.gene.2006.02.023. PMID 16626896.

127. Liu YJ, Hodson MC, Hall BD (2006). "Loss of the flagellum happened only once in the fungal lineage: phylogenetic structure of kingdom Fungi inferred from RNA polymerase II subunit genes". BMC Evolutionary Biology. 6: 74. doi:10.1186/1471-2148-6-74. PMC 1599754 . PMID 17010206.

128. James TY, Letcher PM, Longcore JE, Mozley-Standridge SE, Porter D, Powell MJ, Griffith GW, Vilgalys R (2006). "A molecular phylogeny of the flagellated fungi (Chytridiomycota) and description of a new phylum (Blastocladiomycota)". Mycologia. 98 (6): 860–71. doi:10.3852/mycologia.98.6.860. PMID 17486963.

129. Lockhart RJ, Van Dyke MI, Beadle IR, Humphreys P, McCarthy AJ (August 2006). "Molecular biological detection of anaerobic gut fungi (Neocallimastigales) from landfill sites". Applied and Environmental Microbiology. 72 (8): 5659–61. doi:10.1128/AEM.01057-06. PMC 1538735 . PMID 16885325.

130. Remy W, Taylor TN, Hass H, Kerp H (December 1994). "Four hundred-million-year-old vesicular arbuscular mycorrhizae". Proceedings of the National Academy of Sciences of the United States of America. 91 (25): 11841–3. Bibcode:1994PNAS...9111841R. doi:10.1073/pnas.91.25.11841. PMC 45331 . PMID 11607500.

131. Schüssler A, Schwarzott D, Walker C (2001). "A new fungal phylum, the Glomeromycota: phylogeny and evolution". Mycological Research. 105 (12): 1413–1421. doi:10.1017/S0953756201005196.

132. Alexopoulos et al., p. 145.

133. For an example, see Samuels GJ (February 2006). "Trichoderma: systematics, the sexual state, and ecology". Phytopathology. 96 (2): 195–206. doi:10.1094/PHYTO-96-0195. PMID 18943925.

134. Radford A, Parish JH (June 1997). "The genome and genes of Neurospora crassa". Fungal Genetics and Biology. 21 (3): 258–66. doi:10.1006/fgbi.1997.0979. PMID 9290240.

135. Valverde ME, Paredes-López O, Pataky JK, Guevara-Lara F (January 1995). "Huitlacoche (Ustilago maydis) as a food source--biology, composition, and production". Critical Reviews in Food Science and Nutrition. 35 (3): 191–229. doi:10.1080/10408399509527699. PMID 7632354.

136. Zisova LG (2009). "Malassezia species and seborrheic dermatitis". Folia Medica. 51 (1): 23–33. PMID 19437895.

137. Perfect JR (June 2006). "Cryptococcus neoformans: the yeast that likes it hot". FEMS Yeast Research. 6 (4): 463–8. doi:10.1111/j.1567-1364.2006.00051.x. PMID 16696642.

138. Blackwell M, Spatafora JW (2004). "Fungi and their allies". In Bills GF, Mueller

GM, Foster MS. Biodiversity of Fungi: Inventory and Monitoring Methods. Amsterdam: Elsevier Academic Press. pp. 18–20. ISBN 0-12-509551-1.

139. Gadd GM (January 2007). "Geomycology: biogeochemical transformations of rocks, minerals, metals and radionuclides by fungi, bioweathering and bioremediation". Mycological Research. 111 (Pt 1): 3–49. doi:10.1016/j.mycres.2006.12.001. PMID 17307120.

140. Lindahl BD, Ihrmark K, Boberg J, Trumbore SE, Högberg P, Stenlid J, Finlay RD (2007). "Spatial separation of litter decomposition and mycorrhizal nitrogen uptake in a boreal forest". The New Phytologist. 173 (3): 611–20. doi:10.1111/j.1469-8137.2006.01936.x. PMID 17244056.

141. Barea JM, Pozo MJ, Azcón R, Azcón-Aguilar C (July 2005). "Microbial co-operation in the rhizosphere". Journal of Experimental Botany. 56 (417): 1761–78. doi:10.1093/jxb/eri197. PMID 15911555.

142. Aanen DK (June 2006). "As you reap, so shall you sow: coupling of harvesting and inoculating stabilizes the mutualism between termites and fungi". Biology Letters. 2 (2): 209–12. doi:10.1098/rsbl.2005.0424. PMC 1618886 . PMID 17148364.

143. Nikoh N, Fukatsu T (April 2000). "Interkingdom host jumping underground: phylogenetic analysis of entomoparasitic fungi of the genus cordyceps". Molecular Biology and Evolution. 17 (4): 629–38. doi:10.1093/oxfordjournals.molbev.a026341. PMID 10742053.

144. Perotto S, Bonfante P (December 1997). "Bacterial associations with mycorrhizal fungi: close and distant friends in the rhizosphere". Trends in Microbiology. 5 (12): 496–501. doi:10.1016/S0966-842X(97)01154-2. PMID 9447662.

145. Arnold AE, Mejía LC, Kyllo D, Rojas EI, Maynard Z, Robbins N, Herre EA (December 2003). "Fungal endophytes limit pathogen damage in a tropical tree". Proceedings of the National Academy of Sciences of the United States of America. 100 (26): 15649–54. Bibcode:2003PNAS..10015649A. doi:10.1073/pnas.2533483100. PMC 307622 . PMID 14671327.

146. Paszkowski U (August 2006). "Mutualism and parasitism: the yin and yang of plant symbioses". Current Opinion in Plant Biology. 9 (4): 364–70. doi:10.1016/j.pbi.2006.05.008. PMID 16713732.

147. Hube B (August 2004). "From commensal to pathogen: stage- and tissue-specific gene expression of Candida albicans". Current Opinion in Microbiology. 7 (4): 336–41. doi:10.1016/j.mib.2004.06.003. PMID 15288621.

148. Bonfante P (April 2003). "Plants, mycorrhizal fungi and endobacteria: a dialog among cells and genomes". The Biological Bulletin. 204 (2): 215–20. doi:10.2307/1543562. JSTOR 1543562. PMID 12700157.

149. van der Heijden MG, Streitwolf-Engel R, Riedl R, Siegrist S, Neudecker A, Ineichen K, Boller T, Wiemken A, Sanders IR (2006). "The mycorrhizal contribution to plant productivity, plant nutrition and soil structure in experimental grassland". The New Phytologist. 172 (4): 739–52. doi:10.1111/j.1469-8137.2006.01862.x. PMID 17096799.

150. Selosse MA, Richard F, He X, Simard SW (November 2006). "Mycorrhizal networks: des liaisons dangereuses?". Trends in Ecology & Evolution. 21 (11): 621–8. doi:10.1016/j.tree.2006.07.003. PMID 16843567.
151. Merckx V, Bidartondo MI, Hynson NA (December 2009). "Myco-heterotrophy: when fungi host plants". Annals of Botany. 104 (7): 1255–61. doi:10.1093/aob/mcp235. PMC 2778383 . PMID 19767309.
152. Schulz B, Boyle C (June 2005). "The endophytic continuum". Mycological Research. 109 (Pt 6): 661–86. doi:10.1017/S095375620500273X. PMID 16080390.
153. Clay K, Schardl C (October 2002). "Evolutionary origins and ecological consequences of endophyte symbiosis with grasses". The American Naturalist. 160 Suppl 4 (suppl. 4): S99–S127. doi:10.1086/342161. PMID 18707456.
154. Brodo IM, Sharnoff SD (2001). Lichens of North America. New Haven, Connecticut: Yale University Press. ISBN 0-300-08249-5.
155. Raven PH, Evert RF, Eichhorn, SE (2005). "14—Fungi". Biology of Plants (7 ed.). W. H. Freeman. p. 290. ISBN 978-0-7167-1007-3.
156. Deacon, p. 267.
157. Purvis W (2000). Lichens. Washington, D.C.: Smithsonian Institution Press in association with the Natural History Museum, London. pp. 49–75. ISBN 1-56098-879-7.
158. Kirk et al., p. 378.
159. Deacon, pp. 267–276.
160. Douglas AE (November 1989). "Mycetocyte symbiosis in insects". Biological Reviews of the Cambridge Philosophical Society. 64 (4): 409–34. doi:10.1111/j.1469-185X.1989.tb00682.x. PMID 2696562.
161. Deacon, p. 277.
162. "Entomologists: Brazilian Stingless Bee Must Cultivate Special Type of Fungus to Survive". Sci-News.com. 23 October 2015. Retrieved 25 October 2015.
163. Nguyen NH, Suh SO, Blackwell M (2007). "Five novel Candida species in insect-associated yeast clades isolated from Neuroptera and other insects". Mycologia. 99 (6): 842–58. doi:10.3852/mycologia.99.6.842. PMID 18333508.
164. Filipiak, Michał; Weiner, January (2016-09-01). "Nutritional dynamics during the development of xylophagous beetles related to changes in the stoichiometry of 11 elements". Physiological Entomology: n/a–n/a. doi:10.1111/phen.12168. ISSN 1365-3032.
165. Filipiak, Michał; Sobczyk, Łukasz; Weiner, January (2016-04-09). Fungal Transformation of Tree Stumps into a Suitable Resource for Xylophagous Beetles via Changes in Elemental Ratios. Insects. 7 (2): 13. doi:10.3390/insects7020013.
166. Chandler PJ (2010). A Dipterist's Handbook (2nd Edition). U.K.: The Amateur Entomologists' Society. pp. 1–525.
167. Talbot NJ (2003). On the trail of a cereal killer: Exploring the biology of Magnaporthe grisea. Annual Review of Microbiology. 57: 177–202. doi:10.1146/

annurev.micro.57.030502.090957. PMID 14527276.

168. Paoletti M, Buck KW, Brasier CM (January 2006). Selective acquisition of novel mating type and vegetative incompatibility genes via interspecies gene transfer in the globally invading eukaryote Ophiostoma novo-ulmi. Molecular Ecology. 15 (1): 249–62. doi:10.1111/j.1365-294X.2005.02728.x. PMID 16367844.

169. Gryzenhout M, Wingfield BD, Wingfield MJ (May 2006). New taxonomic concepts for the important forest pathogen Cryphonectria parasitica and related fungi. FEMS Microbiology Letters. 258 (2): 161–72. doi:10.1111/j.1574-6968.2006.00170.x. PMID 16640568.

170. Yang Y, Yang E, An Z, Liu X (May 2007). Evolution of nematode-trapping cells of predatory fungi of the Orbiliaceae based on evidence from rRNA-encoding DNA and multiprotein sequences. Proceedings of the National Academy of Sciences of the United States of America. 104 (20): 8379–84. Bibcode:2007PNAS..104.8379Y. doi:10.1073/pnas.0702770104. PMC 1895958 . PMID 17494736.

171. Nielsen K, Heitman J (2007). Sex and virulence of human pathogenic fungi. Advances in Genetics. Advances in Genetics. 57: 143–73. doi:10.1016/S0065-2660(06)57004-X. ISBN 978-0-12-017657-1. PMID 17352904.

172. Brakhage AA (December 2005). Systemic fungal infections caused by Aspergillus species: epidemiology, infection process and virulence determinants. Current Drug Targets. 6 (8): 875–86. doi:10.2174/138945005774912717. PMID 16375671.

173. Kauffman CA (January 2007). Histoplasmosis: a clinical and laboratory update. Clinical Microbiology Reviews. 20 (1): 115–32. doi:10.1128/CMR.00027-06. PMC 1797635 . PMID 17223625.

174. Cushion MT, Smulian AG, Slaven BE, Sesterhenn T, Arnold J, Staben C, Porollo A, Adamczak R, Meller J (2007). Transcriptome of Pneumocystis carinii during fulminate infection: carbohydrate metabolism and the concept of a compatible parasite. PloS One. 2 (5): e423. Bibcode:2007PLoSO...2..423C. doi:10.1371/journal.pone.0000423. PMC 1855432 . PMID 17487271.

175. Cook GC, Zumla AI (2008). Manson's Tropical Diseases: Expert Consult. Edinburgh, Scotland: Saunders Ltd. p. 347. ISBN 1-4160-4470-1.

176. Simon-Nobbe B, Denk U, Pöll V, Rid R, Breitenbach M (2008). The spectrum of fungal allergy. International Archives of Allergy and Immunology. 145 (1): 58–86. doi:10.1159/000107578. PMID 17709917.

177. Schardl CL, Panaccione DG, Tudzynski P (2006). Ergot alkaloids--biology and molecular biology. The Alkaloids. Chemistry and Biology. The Alkaloids: Chemistry and Biology. 63: 45–86. doi:10.1016/S1099-4831(06)63002-2. ISBN 978-0-12-469563-4. PMID 17133714.

178. van Egmond HP, Schothorst RC, Jonker MA (September 2007). Regulations relating to mycotoxins in food: perspectives in a global and European context. Analytical and Bioanalytical Chemistry. 389 (1): 147–57. doi:10.1007/s00216-007-1317-9. PMID 17508207.

179. Demain AL, Fang A (2000). The natural functions of secondary metabolites. Advances in Biochemical Engineering/Biotechnology. Advances in Biochemical

Engineering/Biotechnology. 69: 1–39. doi:10.1007/3-540-44964-7_1. ISBN 978-3-540-67793-2. PMID 11036689.

180. Rohlfs M, Albert M, Keller NP, Kempken F (October 2007). Secondary chemicals protect mould from fungivory. Biology Letters. 3 (5): 523–5. doi:10.1098/rsbl.2007.0338. PMC 2391202 . PMID 17686752.

181. Molina L, Kahmann R (July 2007). An Ustilago maydis gene involved in H2O2detoxification is required for virulence. The Plant Cell. 19 (7): 2293–309. doi:10.1105/tpc.107.052332. PMC 1955693 . PMID 17616735.

182. Kojic M, Zhou Q, Lisby M, Holloman WK (January 2006). Rec2 interplay with both Brh2 and Rad51 balances recombinational repair in Ustilago maydis. Molecular and Cellular Biology. 26 (2): 678–88. doi:10.1128/MCB.26.2.678-688.2006. PMC 1346908 . PMID 16382157.

183. Michod RE, Bernstein H, Nedelcu AM (May 2008). Adaptive value of sex in microbial pathogens (PDF). Infection, Genetics and Evolution. 8 (3): 267–85. doi:10.1016/j.meegid.2008.01.002. PMID 18295550.

184. Fan W, Kraus PR, Boily MJ, Heitman J (August 2005). Cryptococcus neoformans gene expression during murine macrophage infection. Eukaryotic Cell. 4 (8): 1420–33. doi:10.1128/EC.4.8.1420-1433.2005. PMC 1214536 . PMID 16087747.

185. Lin X, Hull CM, Heitman J (April 2005). Sexual reproduction between partners of the same mating type in Cryptococcus neoformans. Nature. 434 (7036): 1017–21. Bibcode:2005Natur.434.1017L. doi:10.1038/nature03448. PMID 15846346.

186. Fincham JR (March 1989). Transformation in fungi. Microbiological Reviews. 53 (1): 148–70. PMC 372721 . PMID 2651864.

187. Hawkins KM, Smolke CD (September 2008). Production of benzylisoquinoline alkaloids in Saccharomyces cerevisiae. Nature Chemical Biology. 4 (9): 564–73. doi:10.1038/nchembio.105. PMC 2830865 . PMID 18690217.

188. Huang B, Guo J, Yi B, Yu X, Sun L, Chen W (July 2008). Heterologous production of secondary metabolites as pharmaceuticals in Saccharomyces cerevisiae. Biotechnology Letters. 30 (7): 1121–37. doi:10.1007/s10529-008-9663-z. PMID 18512022.

189. Brakhage AA, Spröte P, Al-Abdallah Q, Gehrke A, Plattner H, Tüncher A (2004). Regulation of penicillin biosynthesis in filamentous fungi. Advances in Biochemical Engineering/Biotechnology. 88: 45–90. doi:10.1007/b99257. PMID 15719552.

190. Pan A, Lorenzotti S, Zoncada A (January 2008). Registered and investigational drugs for the treatment of methicillin-resistant Staphylococcus aureus infection. Recent Patents on Anti-Infective Drug Discovery. 3 (1): 10–33. doi:10.2174/157489108783413173. PMID 18221183.

191. Fajardo A, Martínez JL (April 2008). Antibiotics as signals that trigger specific bacterial responses. Current Opinion in Microbiology. 11 (2): 161–7. doi:10.1016/j.mib.2008.02.006. PMID 18373943.

192. Loo DS (2006). Systemic antifungal agents: an update of established and new

therapies. Advances in Dermatology. 22: 101–24. doi:10.1016/j.yadr.2006.07.001. PMID 17249298.

193. Manzoni M, Rollini M (April 2002). Biosynthesis and biotechnological production of statins by filamentous fungi and application of these cholesterol-lowering drugs. Applied Microbiology and Biotechnology. 58 (5): 555–64. doi:10.1007/s00253-002-0932-9. PMID 11956737.

194. Hetland G, Johnson E, Lyberg T, Bernardshaw S, Tryggestad AM, Grinde B (October 2008). Effects of the medicinal mushroom Agaricus blazei Murill on immunity, infection and cancer. Scandinavian Journal of Immunology. 68 (4): 363–70. doi:10.1111/j.1365-3083.2008.02156.x. PMID 18782264.

195. Firenzuoli F, Gori L, Lombardo G (March 2008). The Medicinal MushroomAgaricus blazei Murrill: Review of Literature and Pharmaco-Toxicological Problems. Evidence-Based Complementary and Alternative Medicine. 5 (1): 3–15. doi:10.1093/ecam/nem007. PMC 2249742 . PMID 18317543.

196. Paterson RR (September 2006). Ganoderma - a therapeutic fungal biofactory. Phytochemistry. 67 (18): 1985–2001. doi:10.1016/j.phytochem.2006.07.004. PMID 16905165.

197. Paterson RR (May 2008). Cordyceps: a traditional Chinese medicine and another fungal therapeutic biofactory?. Phytochemistry. 69 (7): 1469–95. doi:10.1016/j.phytochem.2008.01.027. PMID 18343466.

198. el-Mekkawy S, Meselhy MR, Nakamura N, Tezuka Y, Hattori M, Kakiuchi N, Shimotohno K, Kawahata T, Otake T (November 1998). Anti-HIV-1 and anti-HIV-1-protease substances from Ganoderma lucidum. Phytochemistry. 49 (6): 1651–7. doi:10.1016/S0031-9422(98)00254-4. PMID 9862140.

199. El Dine RS, El Halawany AM, Ma CM, Hattori M (June 2008). Anti-HIV-1 protease activity of lanostane triterpenes from the vietnamese mushroom Ganoderma colossum. Journal of Natural Products. 71 (6): 1022–6. doi:10.1021/np8001139. PMID 18547117.

200. Yuen JW, Gohel MD (2005). Anticancer effects of Ganoderma lucidum: a review of scientific evidence. Nutrition and Cancer. 53 (1): 11–7. doi:10.1207/s15327914nc5301_2. PMID 16351502.

201. Sullivan R, Smith JE, Rowan NJ (2006). Medicinal mushrooms and cancer therapy: translating a traditional practice into Western medicine. Perspectives in Biology and Medicine. 49 (2): 159–70. doi:10.1353/pbm.2006.0034. PMID 16702701.

202. Halpern GM, Miller A (2002). Medicinal Mushrooms: Ancient Remedies for Modern Ailments. New York, New York: M. Evans and Co. p. 116. ISBN 0-87131-981-0.

203. Fisher M, Yang LX (2002). Anticancer effects and mechanisms of polysaccharide-K (PSK): implications of cancer immunotherapy. Anticancer Research. 22 (3): 1737–54. PMID 12168863.

204. Kulp K (2000). Handbook of Cereal Science and Technology. CRC Press. ISBN 0-8247-8294-1.

205. Piskur J, Rozpedowska E, Polakova S, Merico A, Compagno C (April 2006). How did Saccharomyces evolve to become a good brewer?. Trends in Genetics. 22 (4): 183–6. doi:10.1016/j.tig.2006.02.002. PMID 16499989.
206. Abe K, Gomi K, Hasegawa F, Machida M (September 2006). Impact of Aspergillus oryzae genomics on industrial production of metabolites. Mycopathologia. 162 (3): 143–53. doi:10.1007/s11046-006-0049-2. PMID 16944282.
207. Hachmeister KA, Fung DY (1993). Tempeh: a mold-modified indigenous fermented food made from soybeans and/or cereal grains. Critical Reviews in Microbiology. 19 (3): 137–88. doi:10.3109/10408419309113527. PMID 8267862.
208. Jørgensen TR (December 2007). Identification and toxigenic potential of the industrially important fungi, Aspergillus oryzae and Aspergillus sojae. Journal of Food Protection. 70 (12): 2916–34. PMID 18095455.
209. O'Donnell K, Cigelnik E, Casper HH (February 1998). Molecular phylogenetic, morphological, and mycotoxin data support reidentification of the Quorn mycoprotein fungus as Fusarium venenatum. Fungal Genetics and Biology. 23 (1): 57–67. doi:10.1006/fgbi.1997.1018. PMID 9501477.
210. Stamets P (2000). Growing Gourmet and Medicinal Mushrooms = [Shokuyō oyobi yakuyō kinoko no saibai]. Berkeley, California: Ten Speed Press. pp. 233–248. ISBN 1-58008-175-4.
211. Hall, pp. 13–26.
212. Kinsella JE, Hwang DH (November 1976). Enzymes of Penicillium roqueforti involved in the biosynthesis of cheese flavor. CRC Critical Reviews in Food Science and Nutrition. 8 (2): 191–228. doi:10.1080/10408397609527222. PMID 21770.
213. Erdogan A, Gurses M, Sert S (August 2003). Isolation of moulds capable of producing mycotoxins from blue mouldy Tulum cheeses produced in Turkey. International Journal of Food Microbiology. 85 (1-2): 83–5. doi:10.1016/S0168-1605(02)00485-3. PMID 12810273.
214. Orr DB, Orr RT (1979). Mushrooms of Western North America. Berkeley, California: University of California Press. p. 17. ISBN 0-520-03656-5.
215. Vetter J (January 1998). Toxins of Amanita phalloides. Toxicon. 36 (1): 13–24. doi:10.1016/S0041-0101(97)00074-3. PMID 9604278.
216. Leathem AM, Dorran TJ (March 2007). Poisoning due to raw Gyromitra esculenta (false morels) west of the Rockies. Cjem. 9 (2): 127–30. doi:10.1017/s1481803500014937. PMID 17391587.
217. Karlson-Stiber C, Persson H (September 2003). Cytotoxic fungi--an overview. Toxicon. 42 (4): 339–49. doi:10.1016/S0041-0101(03)00238-1. PMID 14505933.
218. Michelot D, Melendez-Howell LM (February 2003). Amanita muscaria: chemistry, biology, toxicology, and ethnomycology. Mycological Research. 107 (Pt 2): 131–46. doi:10.1017/S0953756203007305. PMID 12747324.
219. Hall, p. 7.
220. Ammirati JF, McKenny M, Stuntz DE (1987). The New Savory Wild Mushroom.

Seattle, Washington: University of Washington Press. pp. xii–xiii. ISBN 0-295-96480-4.

221. López-Gómez J, Molina-Meyer M (February 2006). The competitive exclusion principle versus biodiversity through competitive segregation and further adaptation to spatial heterogeneities. Theoretical Population Biology. 69 (1): 94–109. doi:10.1016/j.tpb.2005.08.004. PMID 16223517.

222. Becker H (1998). Setting the Stage To Screen Biocontrol Fungi. United States Department of Agriculture, Agricultural Research Service. Retrieved 23 February 2009.

223. Keiller TS. Whey-based fungal microfactory technology for enhanced biological pest management using fungi (PDF). UVM Innovations. Retrieved 29 October 2011.

224. Deshpande MV (1999). Mycopesticide production by fermentation: potential and challenges. Critical Reviews in Microbiology. 25 (3): 229–43. doi:10.1080/10408419991299220. PMID 10524330.

225. Thomas MB, Read AF (May 2007). Can fungal biopesticides control malaria?. Nature Reviews. Microbiology. 5 (5): 377–83. doi:10.1038/nrmicro1638. PMID 17426726.

226. Bush LP, Wilkinson HH, Schardl CL (May 1997). Bioprotective Alkaloids of Grass-Fungal Endophyte Symbioses. Plant Physiology. 114 (1): 1–7. doi:10.1104/pp.114.1.1. PMC 158272 . PMID 12223685.

227. Bouton JH, Latch GCM, Hill NS, Hoveland CS, McCannc MA, Watson RH, Parish JA, Hawkins LL, Thompson FN (2002). Use of nonergot alkaloid-producing endophytes for alleviating tall fescue toxicosis in sheep. Agronomy Journal. 94: 567–574. Review by Parish JA, McCann MA, Watson RH, Hoveland CS, Hawkins LL, Hill NS, Bouton JH (May 2003). Use of nonergot alkaloid-producing endophytes for alleviating tall fescue toxicosis in sheep. Journal of Animal Science. 81 (5): 1316–22. PMID 12772860.

228. Christian V, Shrivastava R, Shukla D, Modi HA, Vyas BR (April 2005). Degradation of xenobiotic compounds by lignin-degrading white-rot fungi: enzymology and mechanisms involved. Indian Journal of Experimental Biology. 43 (4): 301–12. PMID 15875713.

229. Fungi to fight ‹toxic war zones'. BBC News. 5 May 2008.

230. Fomina M, Charnock JM, Hillier S, Alvarez R, Gadd GM (July 2007). Fungal transformations of uranium oxides. Environmental Microbiology. 9 (7): 1696–710. doi:10.1111/j.1462-2920.2007.01288.x. PMID 17564604.

231. Fomina M, Charnock JM, Hillier S, Alvarez R, Livens F, Gadd GM (May 2008). Role of fungi in the biogeochemical fate of depleted uranium. Current Biology. 18 (9): R375–7. doi:10.1016/j.cub.2008.03.011. PMID 18460315.

232. Beadle GW, Tatum EL (November 1941). Genetic Control of Biochemical Reactions in Neurospora. Proceedings of the National Academy of Sciences of the United States of America. 27 (11): 499–506. Bibcode:1941PNAS...27..499B. doi:10.1073/pnas.27.11.499. PMC 1078370 . PMID 16588492.

233. Datta A, Ganesan K, Natarajan K (1989). Current trends in Candida albicans research. Advances in Microbial Physiology. Advances in Microbial Physiology. 30: 53–88. doi:10.1016/S0065-2911(08)60110-1. ISBN 978-0-12-027730-8. PMID 2700541.

234. Dean RA, Talbot NJ, Ebbole DJ, Farman ML, Mitchell TK, Orbach MJ, et al. (April 2005). The genome sequence of the rice blast fungus Magnaporthe grisea. Nature. 434 (7036): 980–6. Bibcode:2005Natur.434..980D. doi:10.1038/nature03449. PMID 15846337.

235. Daly R, Hearn MT (2005). Expression of heterologous proteins in Pichia pastoris: a useful experimental tool in protein engineering and production. Journal of Molecular Recognition. 18 (2): 119–38. doi:10.1002/jmr.687. PMID 15565717.

236. Schlegel HG (1993). General Microbiology. Cambridge, UK: Cambridge University Press. p. 360. ISBN 0-521-43980-9.

237. Joseph B, Ramteke PW, Thomas G (2008). Cold active microbial lipases: some hot issues and recent developments. Biotechnology Advances. 26 (5): 457–70. doi:10.1016/j.biotechadv.2008.05.003. PMID 18571355.

238. Kumar R, Singh S, Singh OV (May 2008). Bioconversion of lignocellulosic biomass: biochemical and molecular perspectives. Journal of Industrial Microbiology & Biotechnology. 35 (5): 377–91. doi:10.1007/s10295-008-0327-8. PMID 18338189.

239. Trichoderma spp., including T. harzianum, T. viride, T. koningii, T. hamatum and other spp. Deuteromycetes, Moniliales (asexual classification system). Biological Control: A Guide to Natural Enemies in North America. Retrieved 10 July 2007.

240. Olempska-Beer ZS, Merker RI, Ditto MD, DiNovi MJ (July 2006). Food-processing enzymes from recombinant microorganisms--a review. Regulatory Toxicology and Pharmacology. 45 (2): 144–158. doi:10.1016/j.yrtph.2006.05.001. PMID 16769167.

241. Polizeli ML, Rizzatti AC, Monti R, Terenzi HF, Jorge JA, Amorim DS (June 2005). Xylanases from fungi: properties and industrial applications. Applied Microbiology and Biotechnology. 67 (5): 577–91. doi:10.1007/s00253-005-1904-7. PMID 15944805.

Index